International Harvester Company: Building Plans and Specifications

A compilation of the Original Plans

From International Harvester Company

Which were made available from 1911 to 1914

All images in this publication from Wisconsin Historical Society (WHS) and are noted with the corresponding image number in caption and are used with the written permission of the WHS, wisconsinhistory.org in conjunction with Navistar, inc.
Warehouse expansion and any other images are from the private collection of myself, Sarah Tomac, and are noted as such.

All images are historical of nature.
These plans are not approved for commercial or residential building use. The content of this book are for historical use and do not reflect current scientific knowledge, policies or practices. This publication is for entertainment and educational purposes only.

1831 Press, LLC

IBSN: 978-1-952265-06-8

Old is New Again

Kit homes.
We've all heard about them, companies most famously known are Sears, Roebuck and Co., Montgomery Ward, William A. Radford, to name a few. These homes were designed with affordability in mind, quality materials, and to help lumber yards, window and door companies sell the dream of a home with a garden, or on a town lot to many Americans.

These plans reflected the way of living in their day, with the majority of the plans being offered from early 1900s to the 1950s. Looking through the decades of these plans show the changes throughout our way of life. Plans of this nature are still available today, while not as prolific, many builders use these plans as a base for new home construction, with many home owners customizing these plans to suit their taste and way of living.
Many of these plans are designed for city life, some designed for narrow lots and in the early 1900s, room for hired help as well.

How many of these houses were specifically designed with a farm life in mind? The short answer, not many. The farm houses of these days often were less ornate in design details, as many homes could not afford the extra cost associated with fancier scroll work and elaborate trims. Farms that featured a skilled carpenter are still evident in the century old buildings today.

When observing the plans throughout the decades, look at the way kitchens, bathrooms (if there are any!) and closets change in the interior. The exterior of the homes change with the trends as well, from ornate Victorian facades, to Craftsman to Mid-Century Designs. Many other styles of buildings are available as well, I encourage you to look around your area to identify them and the time period they were built. International Harvester introduced their plans in 1912, and feature simple straight cuts that most any carpenter could master. The time period of this design is classified as a transitional style, or early craftsman. The fancy fretwork found on Victorian homes is reduced to the details in the change of siding, and the interior shows a style that soon became known as craftsman, defined to showcase the warmth of the wood and grains through stains instead of paint.

What about the Homestead?
Many needed outbuildings, from the barns to a powerhouse to an ice house, there were many needs that a successful farm or ranch often looked like a small town. Some places had a list of buildings from a Blacksmith shop to a Water tower, and every in-between. The McCormick Homestead in West Virginia where Cyrus McCormick grew up features a smokehouse, a schoolroom and a teachers residence, all built with bricks. There was also a spring-house, as well as the grist mill and the well known blacksmith shop. Chicken coops, Horse stables, cattle barn, piggery and servant quarters are just some of the buildings that were on the site as well as the main house.
My home farm still features many buildings, some of which have been torn down and rebuilt over the years. Every farm over the years has learned to diversify and grow with the changes of invention and progression. To hold tight to these early ways of life without recognizing the advancement of mankind and adapting spells a certain end to the way we know. We will always need to eat, and farms will always be the strength of any country, wanted or not. Without them, we would not be clothed, fed, or in a home.

Enjoy these plans that International Harvester provided to anyone who wrote to them asking for help with developing their farm into a prosperous and efficient business of producing food to feed more than their own family, they feed the world.

Sarah

Contents

Introduction found in the front of all building plans from International Harvester, 1912

"These plans and specifications are the work of experts and you will find them reliable. There may be some conditions on your farm which will necessitate changes in these plans. If such is the case, we suggest that you write the I H C Service Bureau, giving as complete informations as possible.

A pencil sketch showing the arrangement of other buildings, pastures, and fences, will be of assistance to us in making alterations in case plans do not conform with your specific conditions. If you wish further detailed information, the Service Bureau will take pleasure in giving you all the assistance possible.

This Service Bureau which is co-operating with the State Agricultural Colleges, Experiment Stations, and the Department of Agriculture, is also at your service in answering any questions pertaining to farm crops, soils, land drainage, farm mechanics, and other problems relative to agriculture that may be confronting you.

Address all requests for information to the
I H C Service Bureau, Harvester Building, Chicago, U.S.A."

Suggestions For Building a Farm House

The plans presented herewith provide for a substantial, attractive, modern, seven-room farm dwelling. In the past more money, time, and pains have usually been spent on building good barns and other farm buildings than on the farm home. Today, farmers are giving more attention to the house, not to the detriment of other farm buildings, but in addition to the care and money they spend on these.

In the plans given herewith particular attention has been directed toward making the house adaptable to farm conditions. For instance, the ground floor is so arranged that you can enter from the screened porch on the rear into the hallway and go directly up stairs without passing through any other room of the house. This is a desirable feature where the sleeping quarters for the farm help are in the house. At the same time, it is possible to come in the front entrance and go up the stairs without passing through any other part of the house. The living and dining room entrance is from this main hall, so that to enter or leave the dining room it is not necessary to pass through the kitchen or through the parlor.

On the second floor are four commodious bedrooms, all opening on to the center hall, with access to the bath. From the side bedroom on this floor is an entrance to the second story sleeping porch. This is a convenience which many farmers have overlooked. The joy and benefits of outdoor sleeping in summer are so great that every farm home should have a sleeping porch.

This house, as laid out, will cost approximately $2,700 to build, based on the price of material and labor in Illinois. The price will vary in localities, depending on the freight rates on material from the source of supply, and local labor conditions. The figure can be reduced by alterations in construction and elimination of such features as are not found necessary. For instance, brick might be used instead of rubble stone, which might save 10 per cent on the cost of masonry. You will note from the plans that a basement is provided under the entire house. The basement under the front and side wings might be eliminated, providing only a small basement under the central portion of the house. Board floors can be substituted instead of cement floors. Instead of clapboards or beveled siding, drop siding might be used. This will effect some saving, although it will not make as attractive an exterior as the original specifications.

Some saving might be effected by reducing the dimensions of this house; that is, two feet might be cut off the front of the living room, and perhaps two feet from the side of the parlor. It will not be advisable to make many further cuts, owing to the fact that the bedrooms on the second floor will be too small to be desirable.

The interior trimming follows the modern tendency of plain, substantial, attractive woodwork. There are no fancy scrolls and no expensive cabinet work indicated. In the pantry are drawers and cupboards, shown in detail on Sheet 9, which, while not expensive can be replaced by plain shelving if it is found desirable to make the change. You will note from the details shown on Sheet 9 that the stairway is simple though attractive.

It is practically impossible for us to give complete specifications and material list. If you desir to know just what these plans will cost you, the most satisfactory way would be to take them to your contractor and ask him to give you a rough figure. You might, at the same time call attention to such features as you think could be altered to make the plans more nearly meet the price you wish to put into your house.

This is one of two house plans that Harvester had available, the second one was called a farm dwelling.

The following pages are of the Farm House No.1

Interesting aspects include an upstairs only bathroom, a sleeping porch, small closets, and the cistern.

Right: Cover of the Building Plan booklet, WHS147942
Left: Introduction to building a Farm House, Tomac Collection

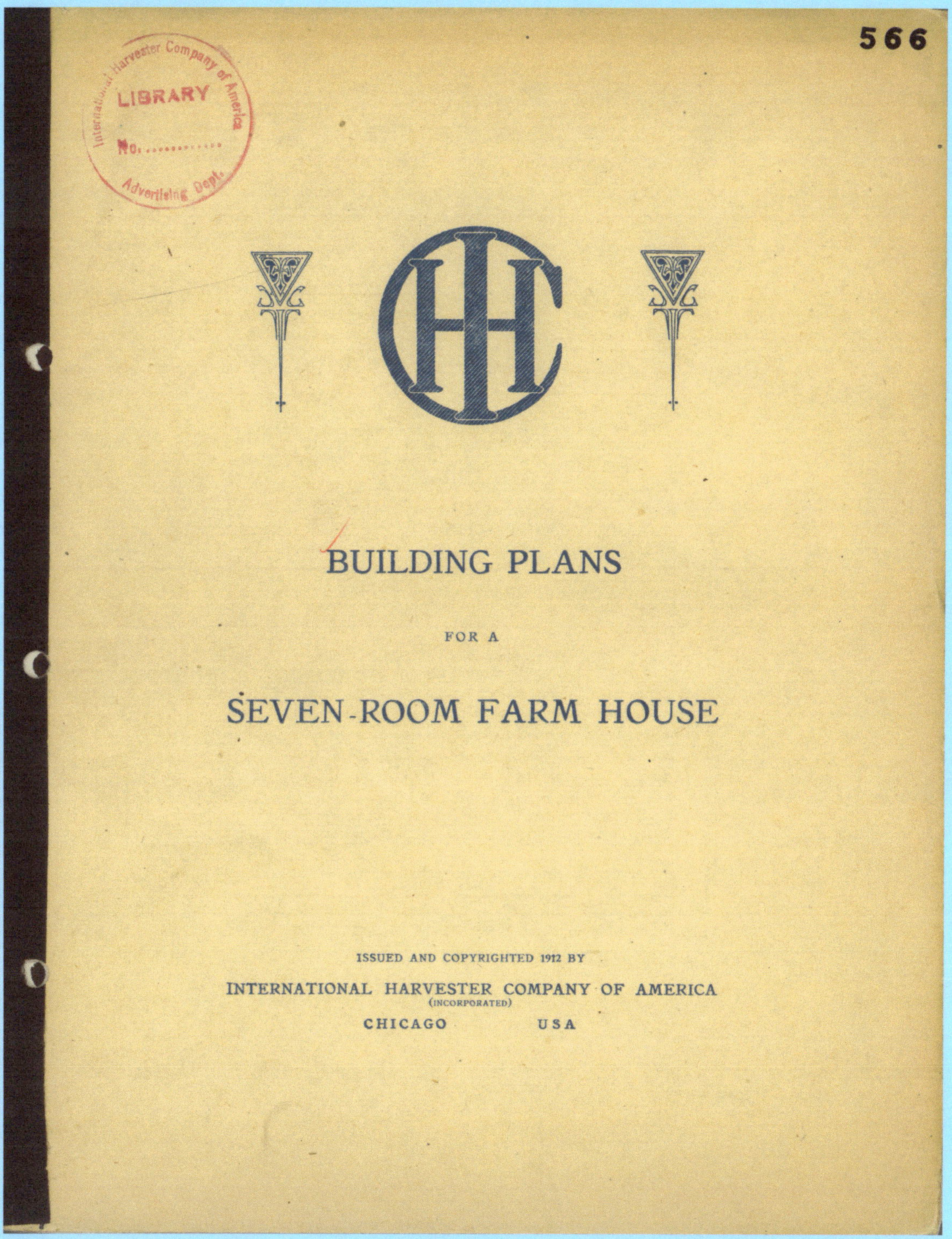

BUILDING PLANS

FOR A

SEVEN-ROOM FARM HOUSE

ISSUED AND COPYRIGHTED 1912 BY
INTERNATIONAL HARVESTER COMPANY OF AMERICA
(INCORPORATED)
CHICAGO USA

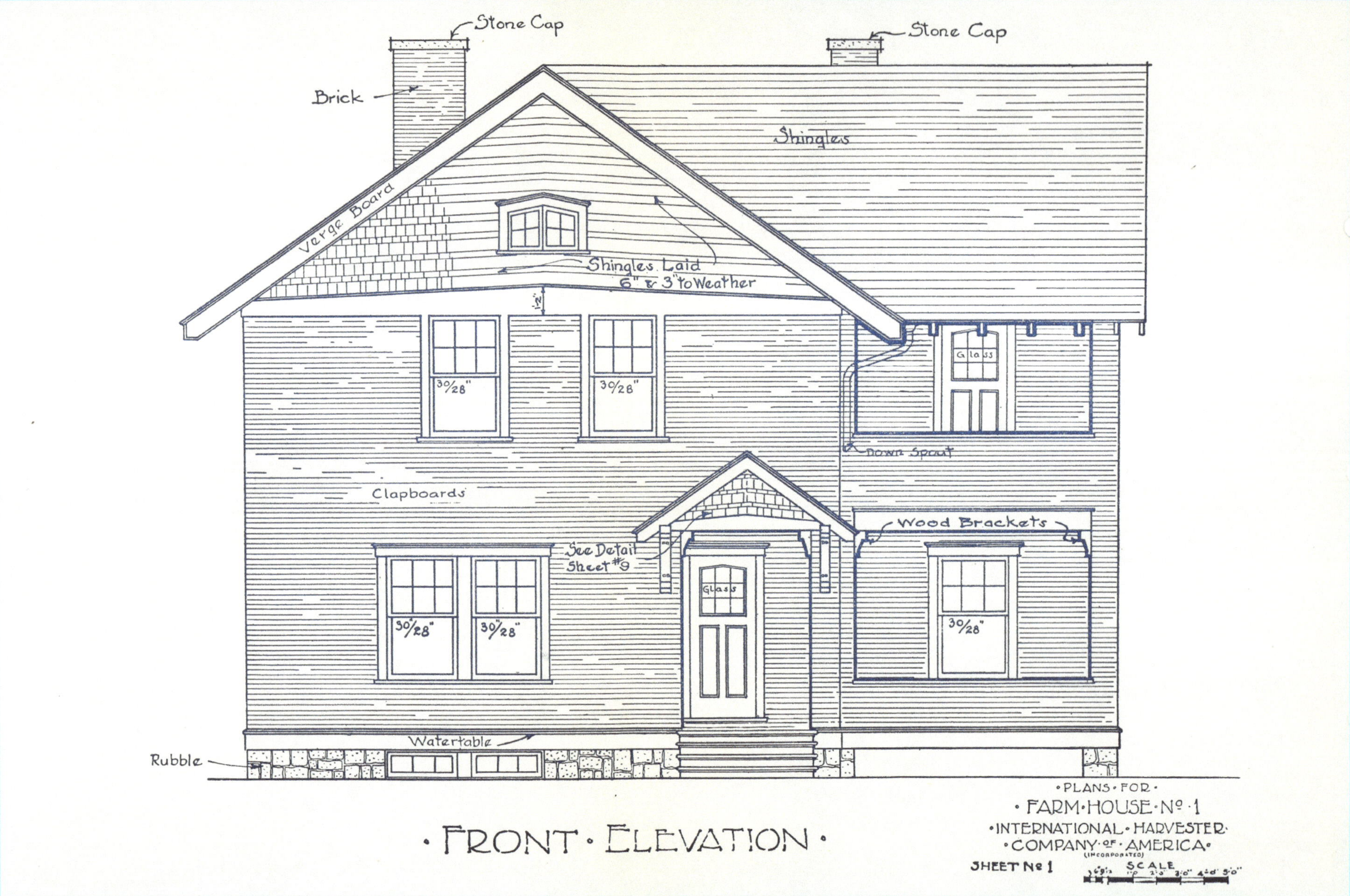

Front Elevation, WHS 147501

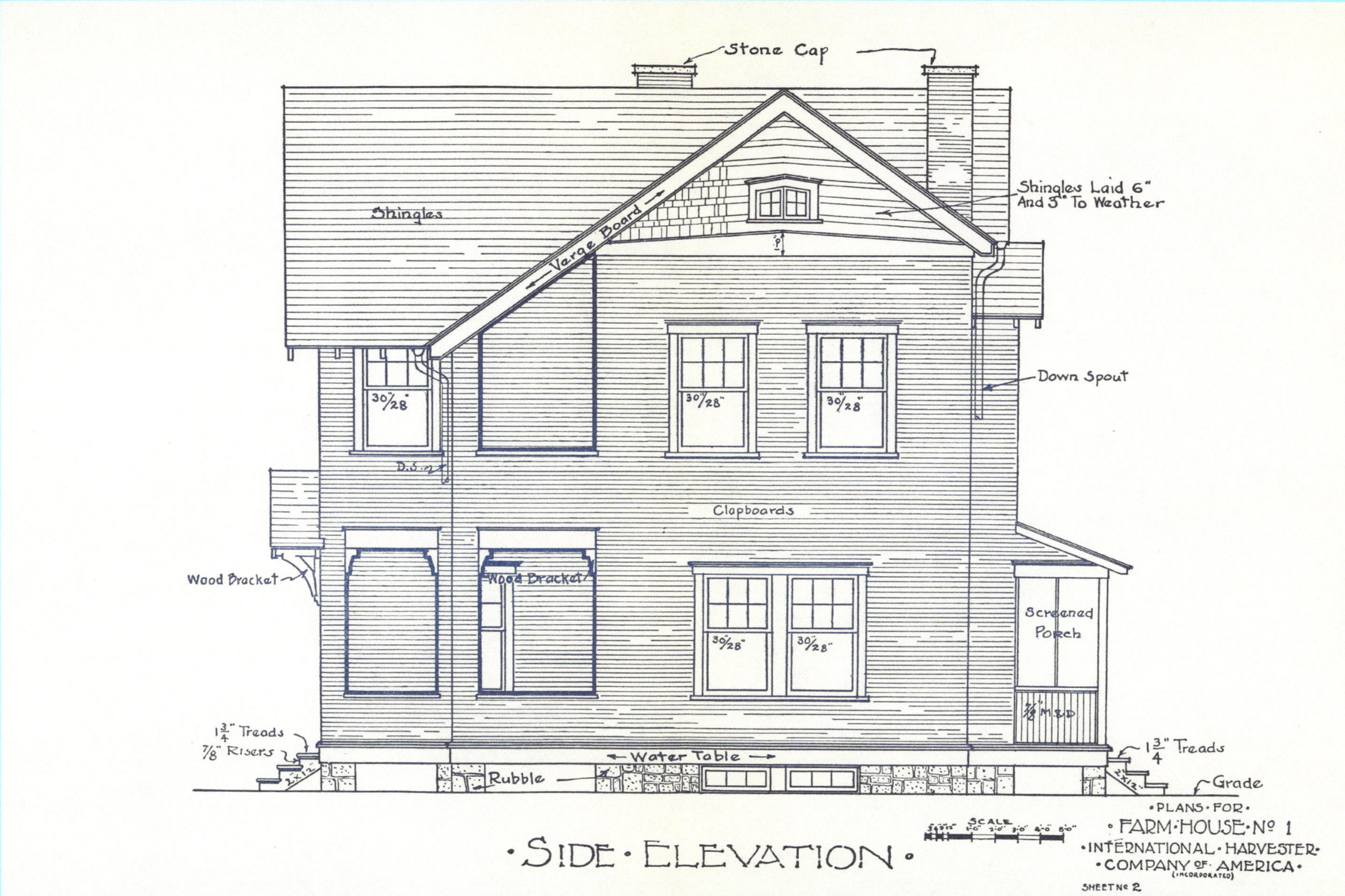

Side Elevation WHS 150169

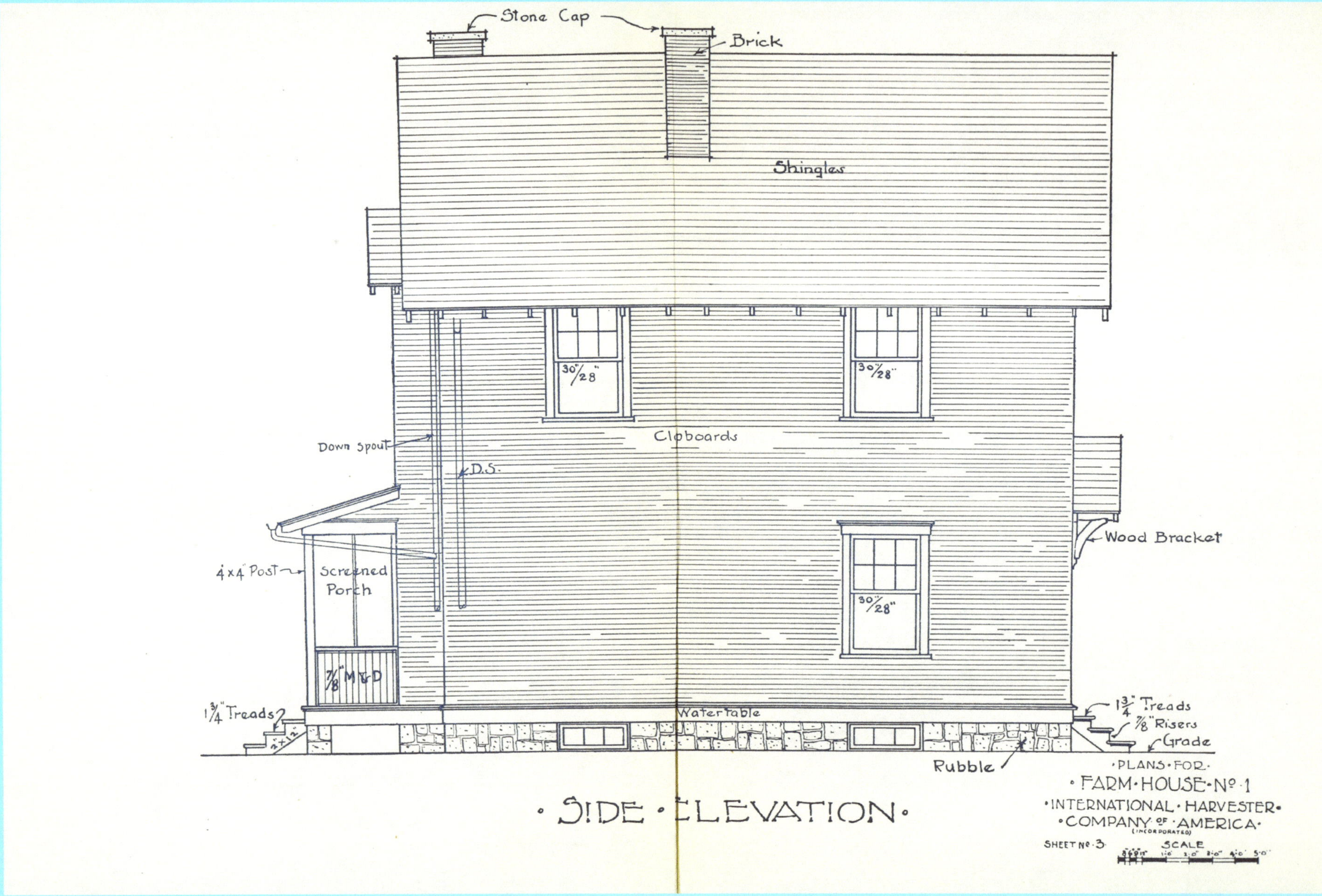

Side Elevation WHS150170

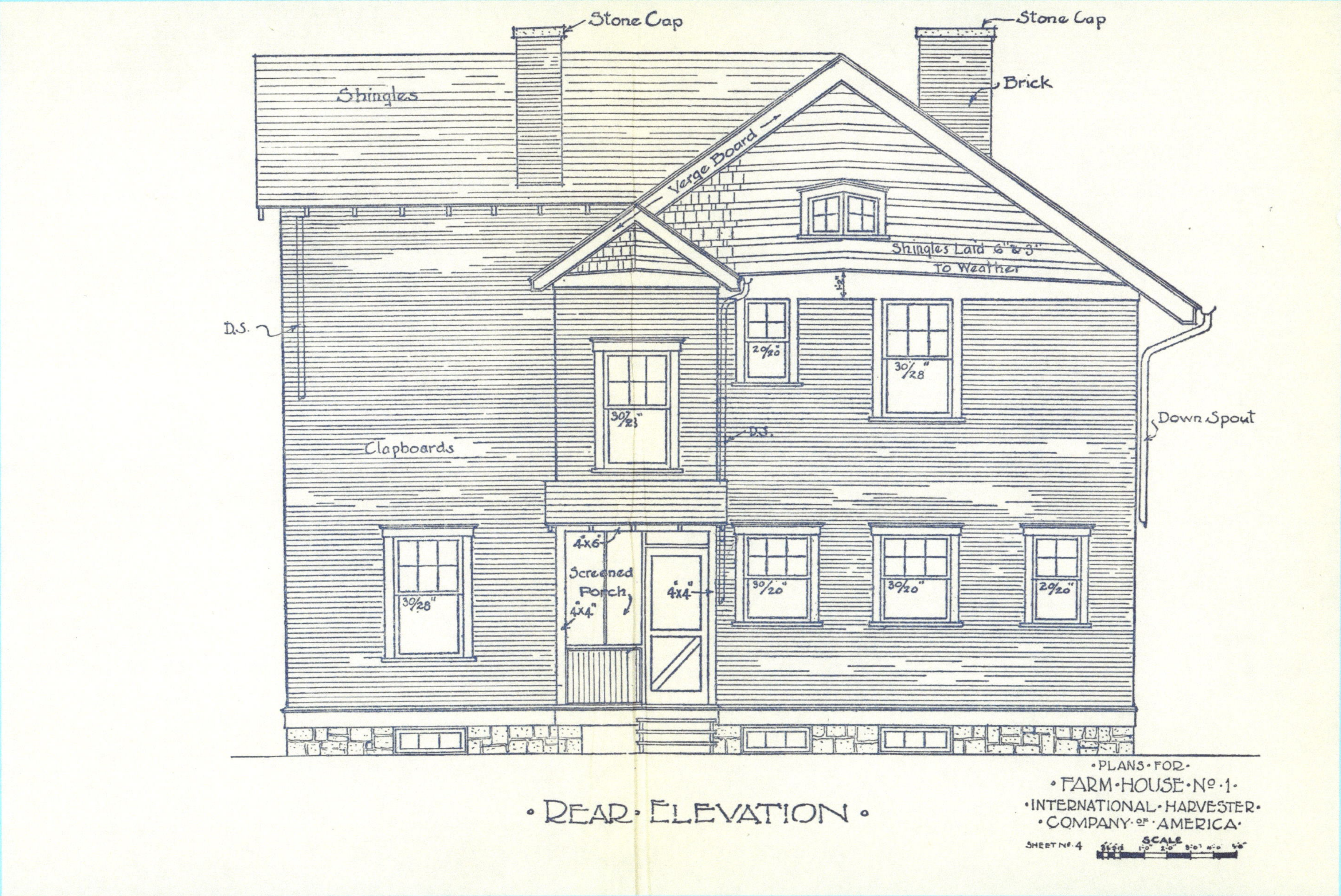

Rear Elevation WHS150174

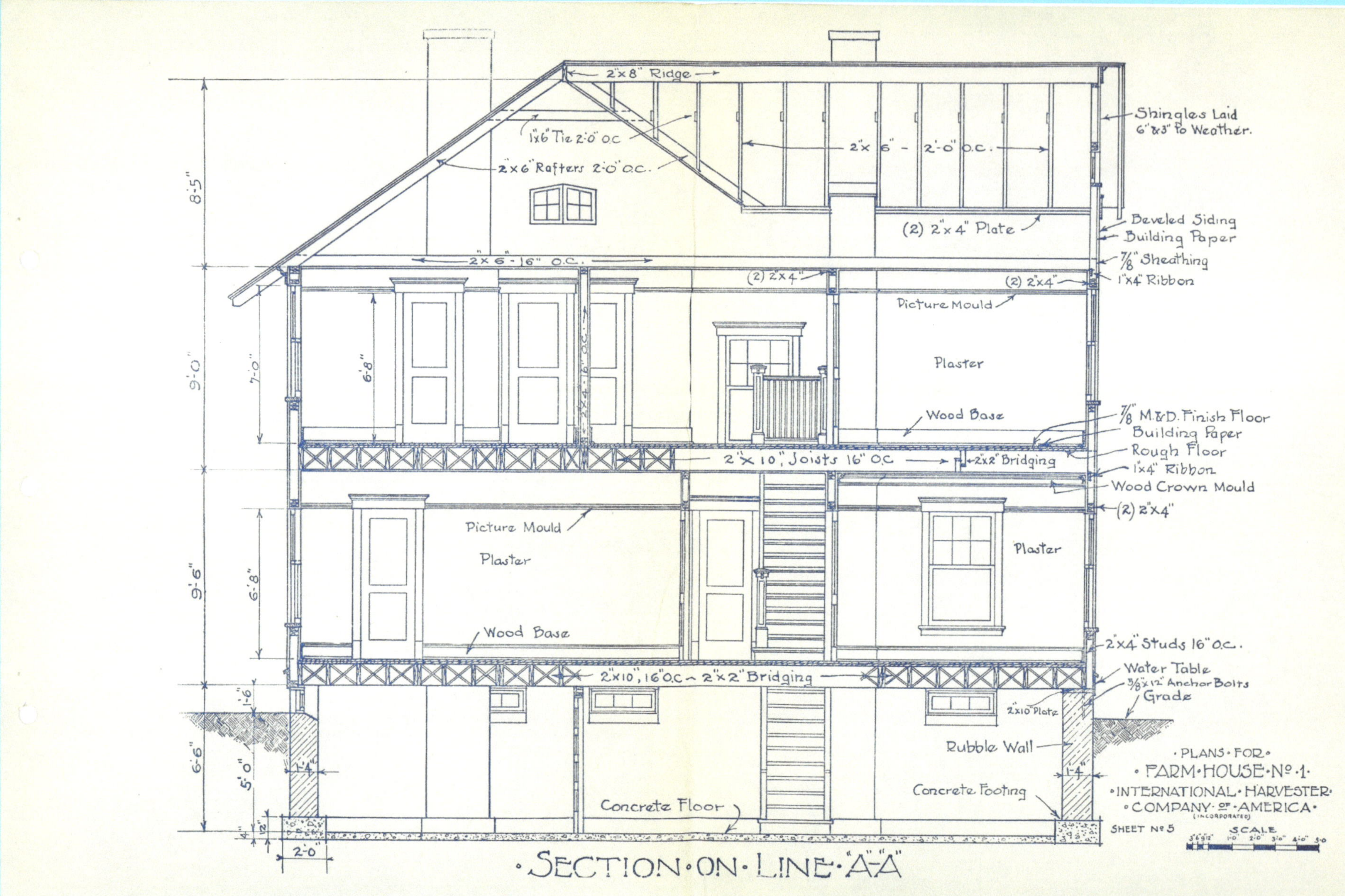

Cross Section Showing trim details. WHS150173

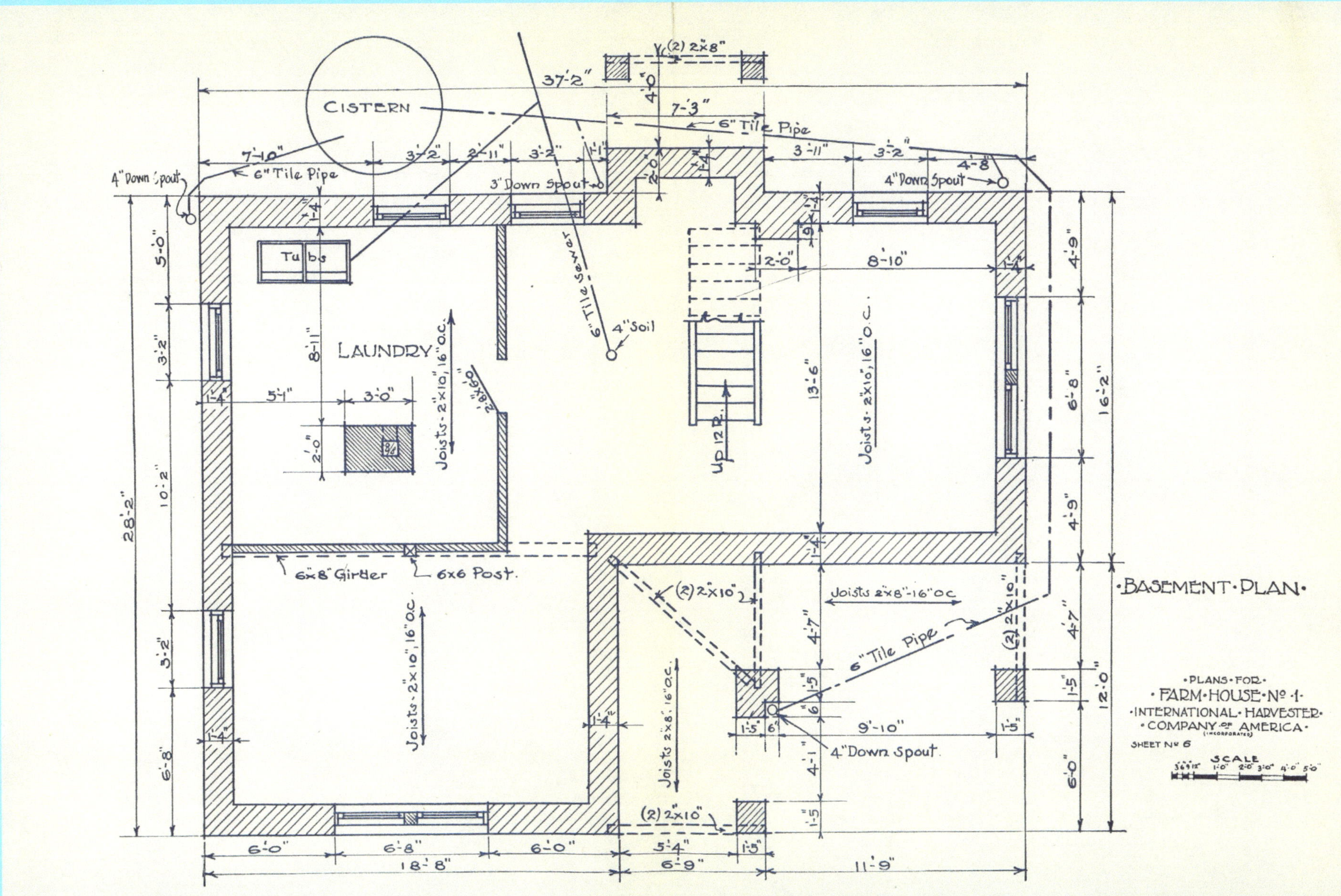

Basement Plan WHS150174

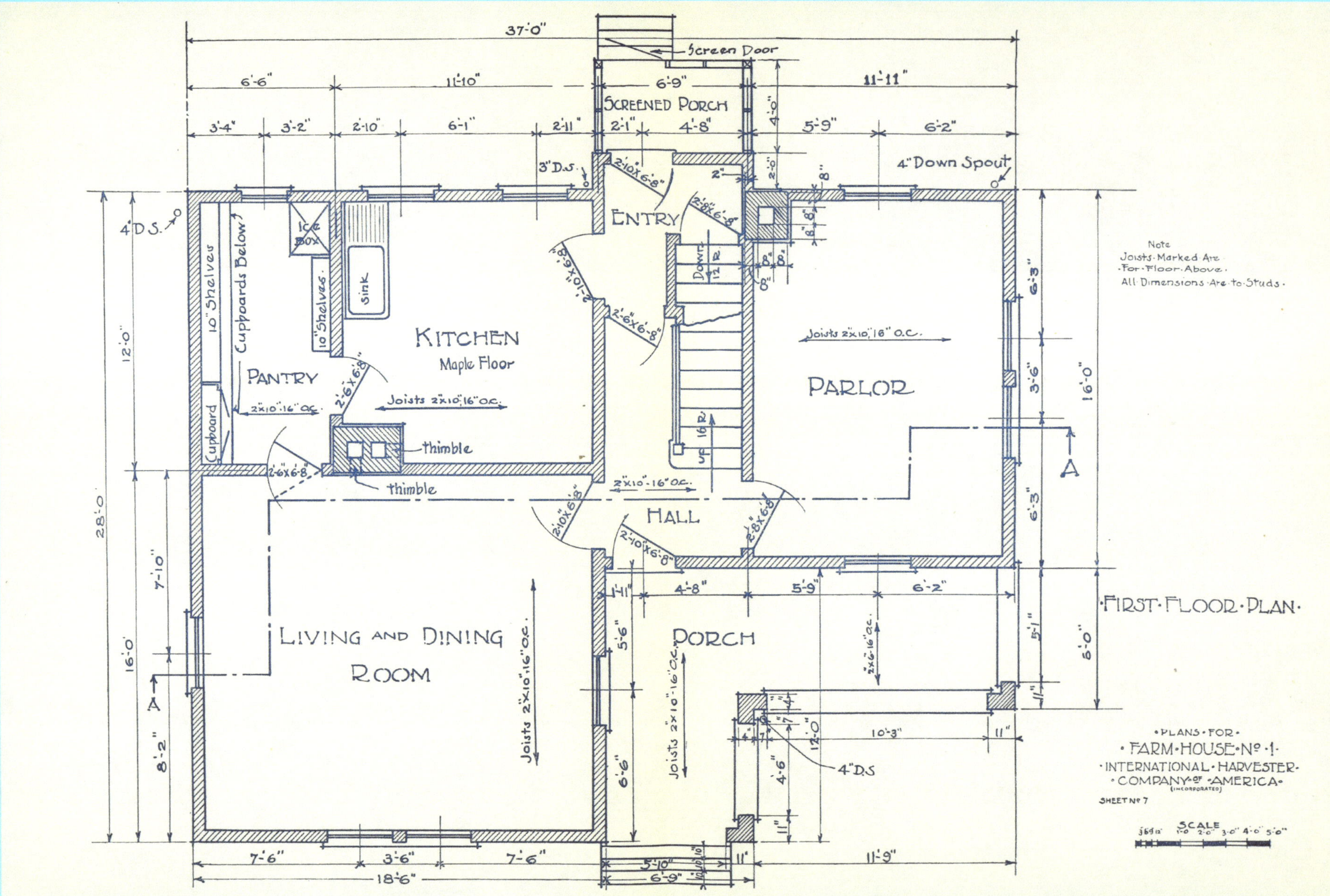

First Floor Plan WHS150176

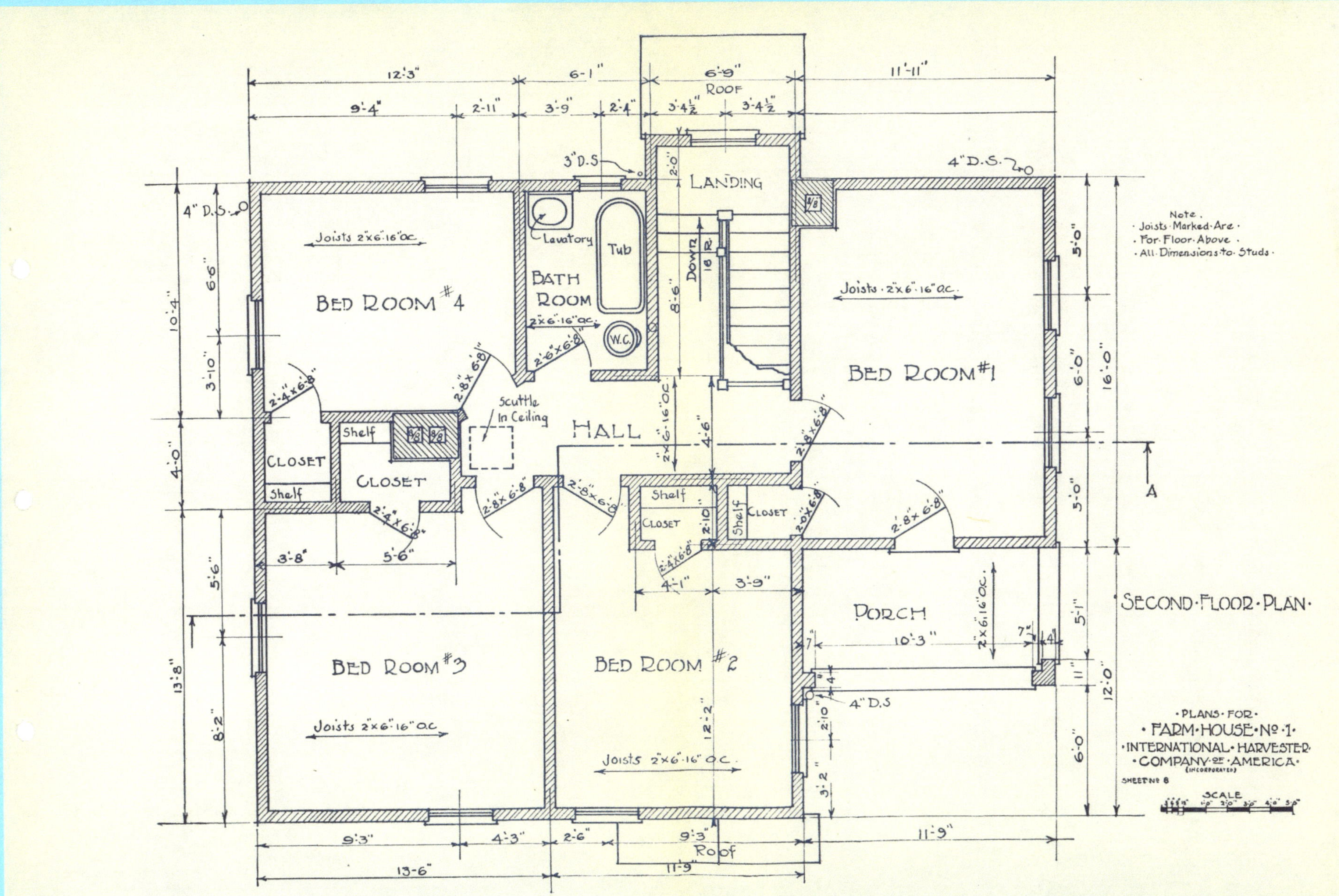

Second Floor Plan WHS150177

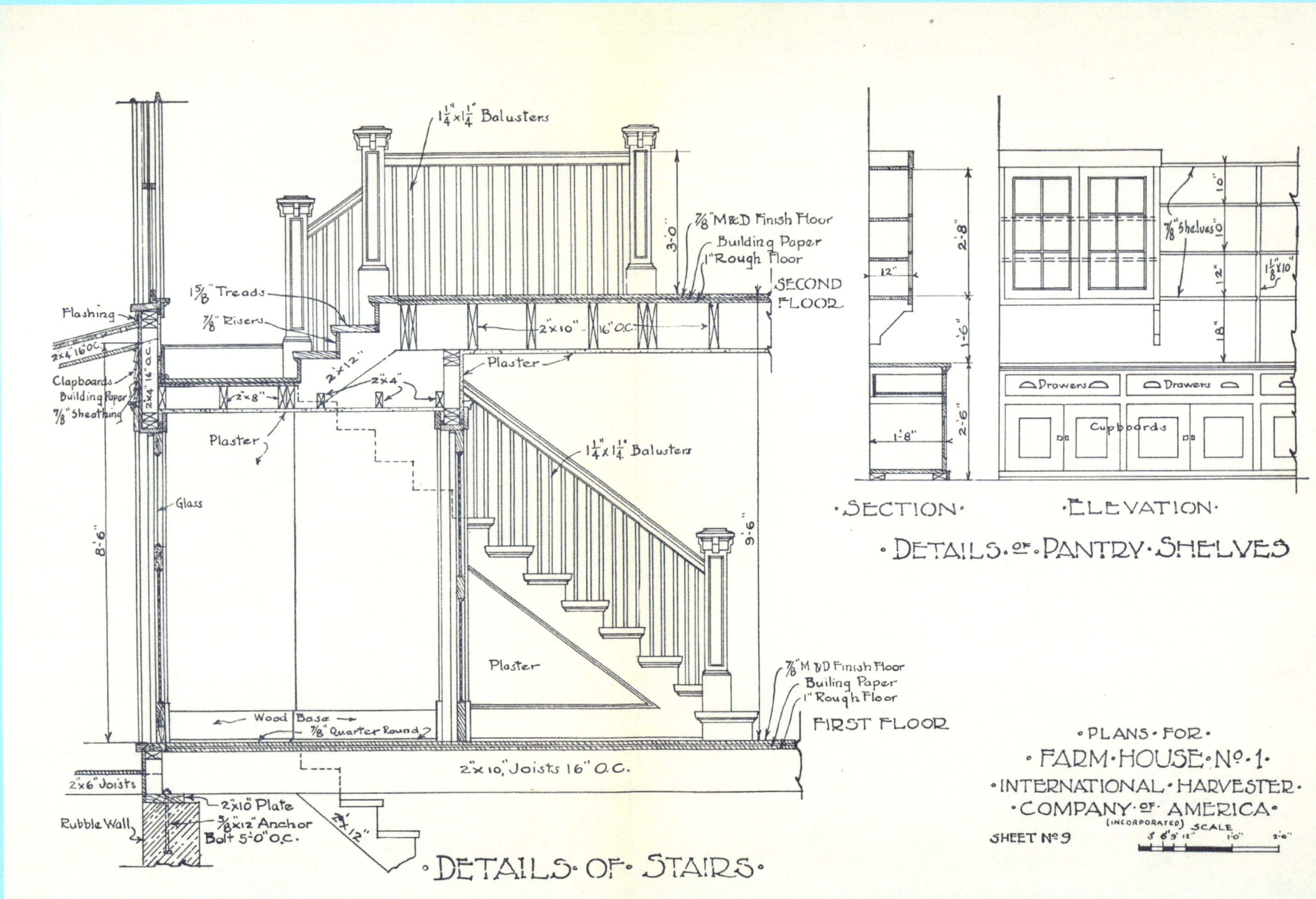

Details of Stairs and Pantry Shelves. WHS150178

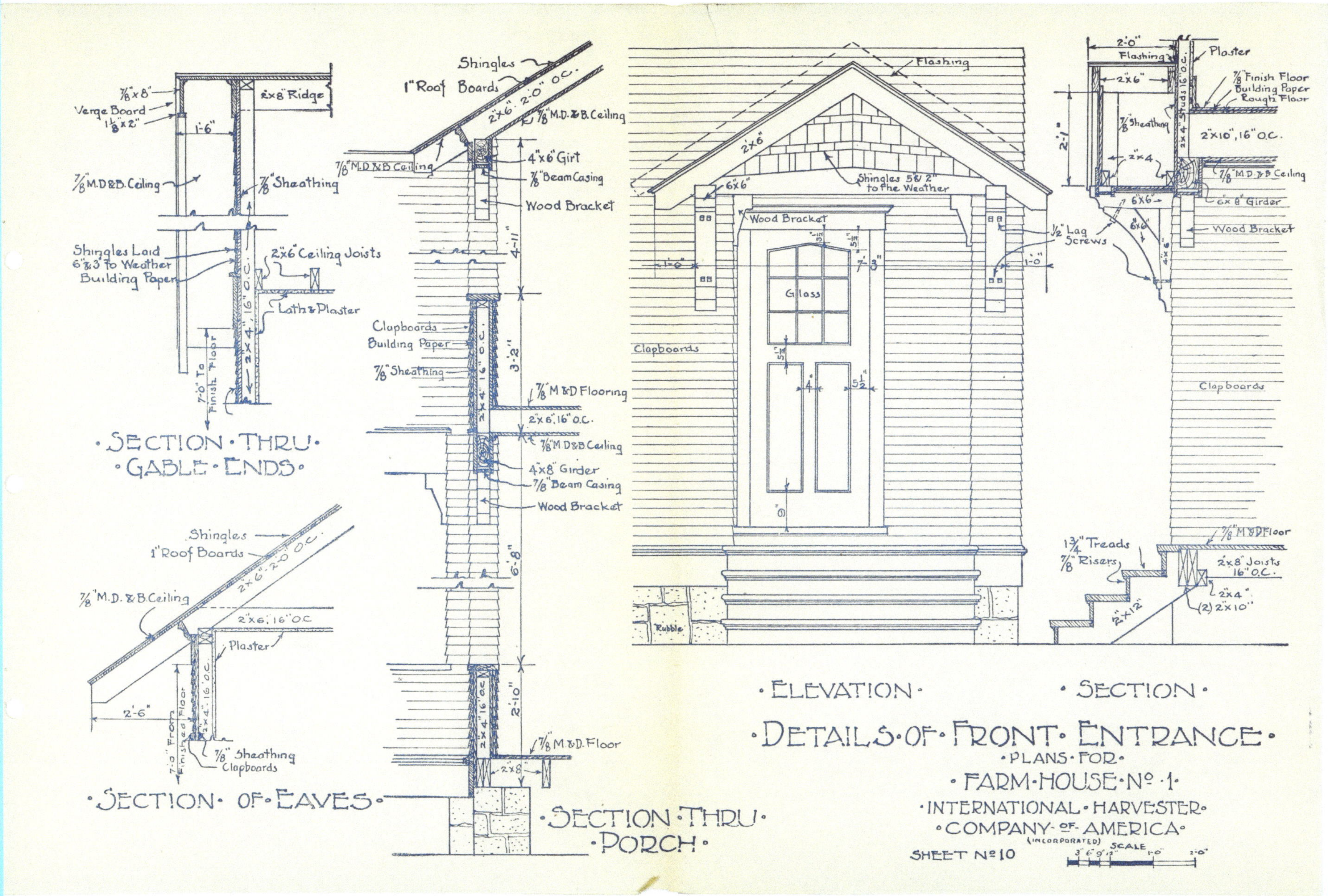

Details of Front Entrance, Section of Eaves, Entrance. WHS150180

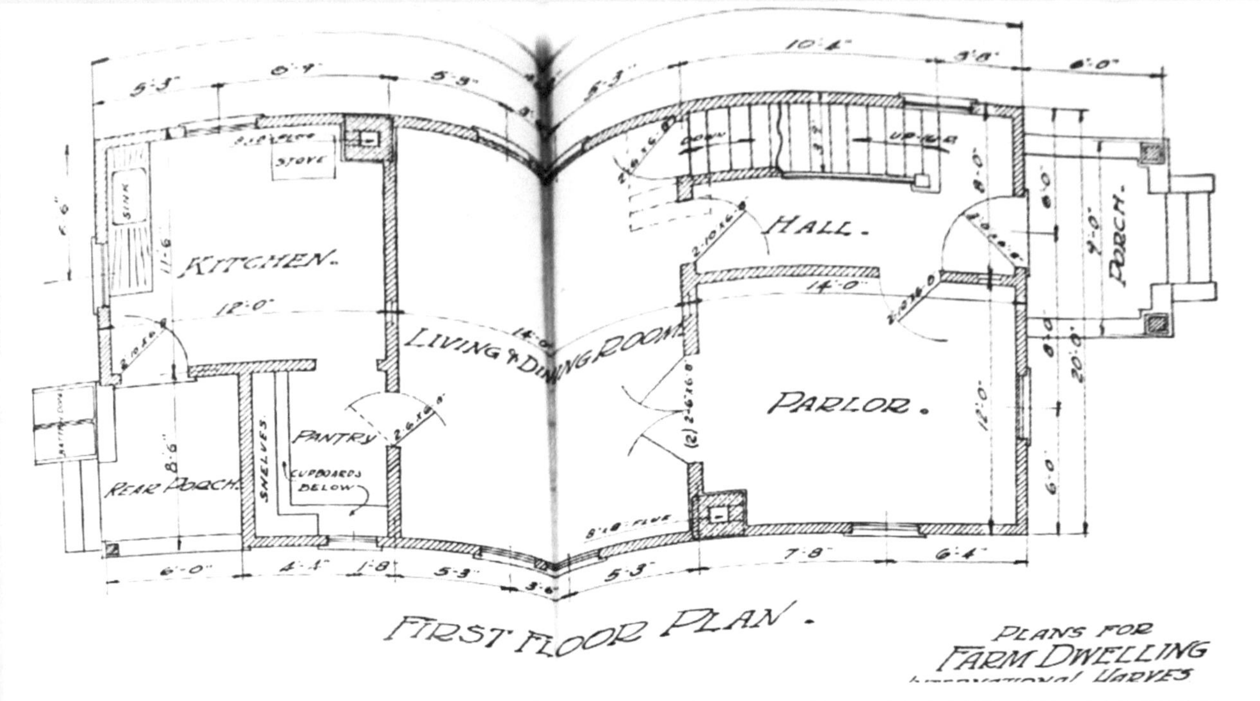
KITCHEN.
STOVE
SINK
LIVING & DINING ROOM.
PANTRY
SHELVES
CUPBOARDS BELOW
REAR PORCH.
HALL.
PARLOR.
PORCH.
FIRST FLOOR PLAN.
PLANS FOR
FARM DWELLING

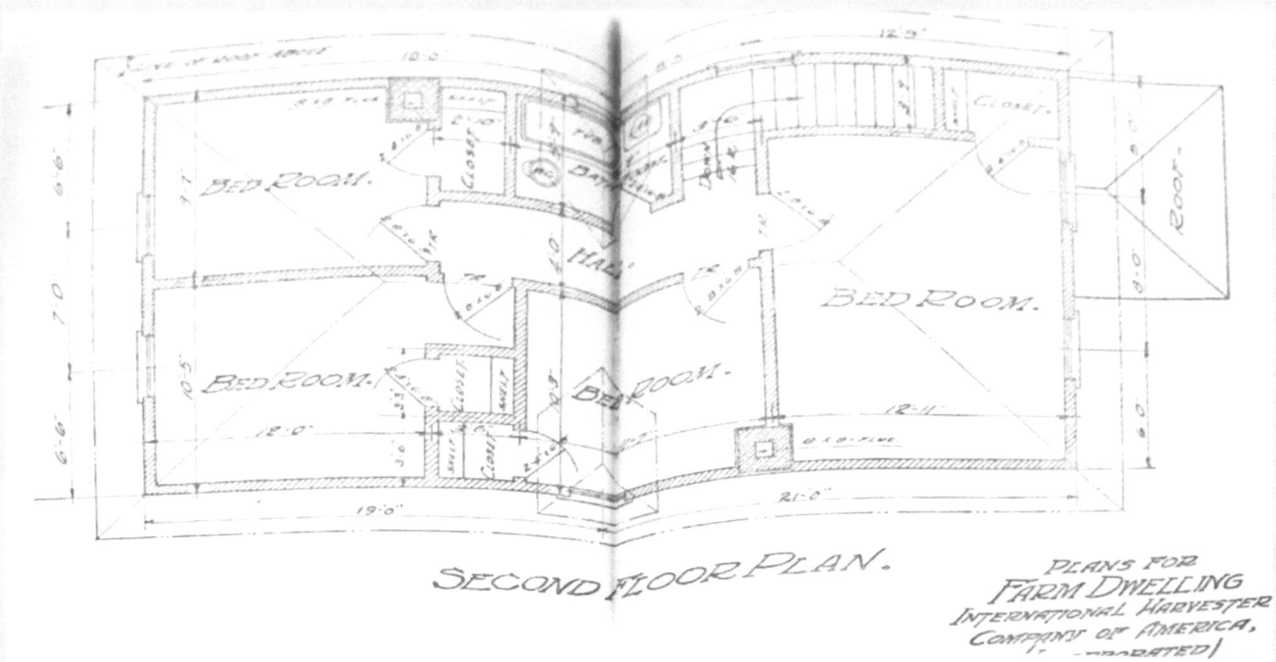
BED ROOM.
BED ROOM.
BED ROOM.
BED ROOM.
HALL
CLOSET
ROOF.
SECOND FLOOR PLAN.
PLANS FOR
FARM DWELLING
INTERNATIONAL HARVESTER
COMPANY OF AMERICA,

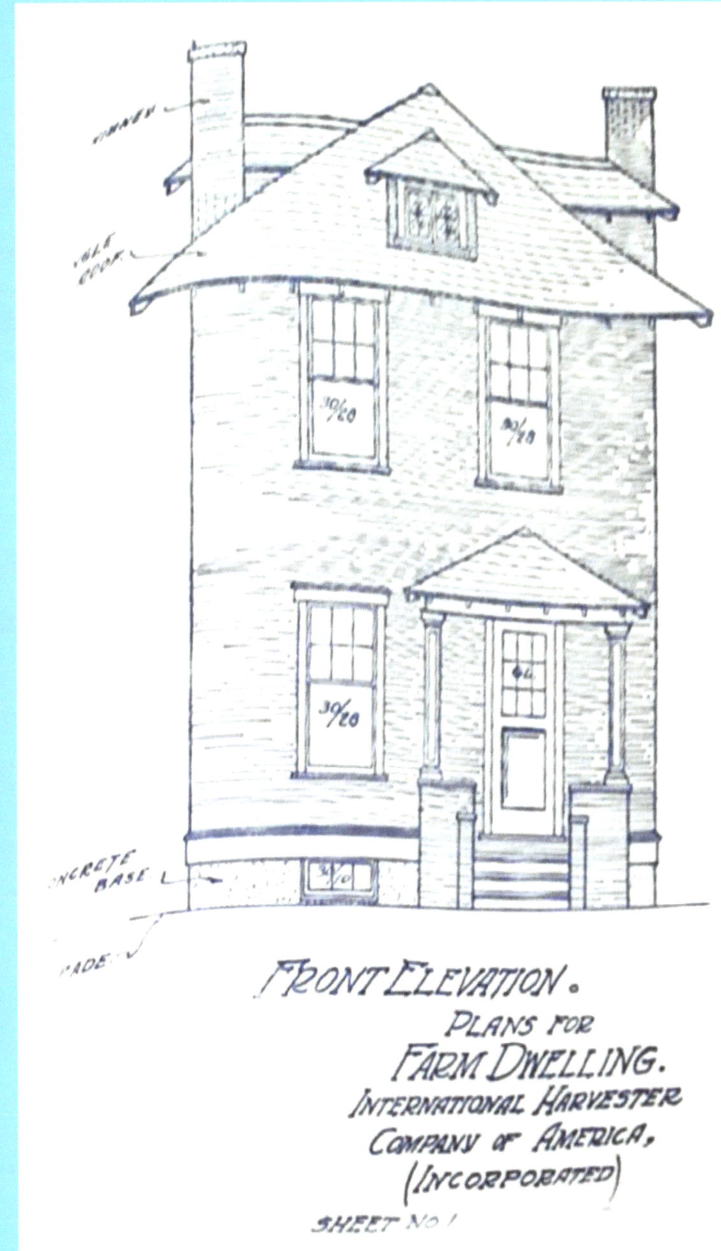

These Photos are of the Farm Dwelling that International Harvester also provided as an option for building.
This home was better suited for a narrow area, or for tucking among trees, as it is designed.
The 20 foot wide and 40 foot deep house featured two small porches, with a kitchen in the rear. The second plan also featured two stories, with the four bedrooms and bathroom located upstairs.
A full basement was also designed for this plan as well.

(These Photos are re-touched from original images found elsewhere, at this time, they are not known to be in the WHS archives)

The General Farm Barn was designed for anyone with a basic level of carpentry skill to build. The era required farms and ranches to be multi-talented, which is still a requirement today.

Traditional Gable or Hip roof barns were and still are common place on many farms, as well as many other roof lines.

The following pages show great details for dimensions, layout and size. Designed to easily be created larger or smaller, to accommodate any farm.

WHS147496

BUILDING PLANS
SPECIFICATIONS
AND
MATERIAL LIST
FOR A
GENERAL FARM BARN

ISSUED AND COPYRIGHTED 1912 BY

INTERNATIONAL HARVESTER COMPANY OF AMERICA
(INCORPORATED)

CHICAGO USA

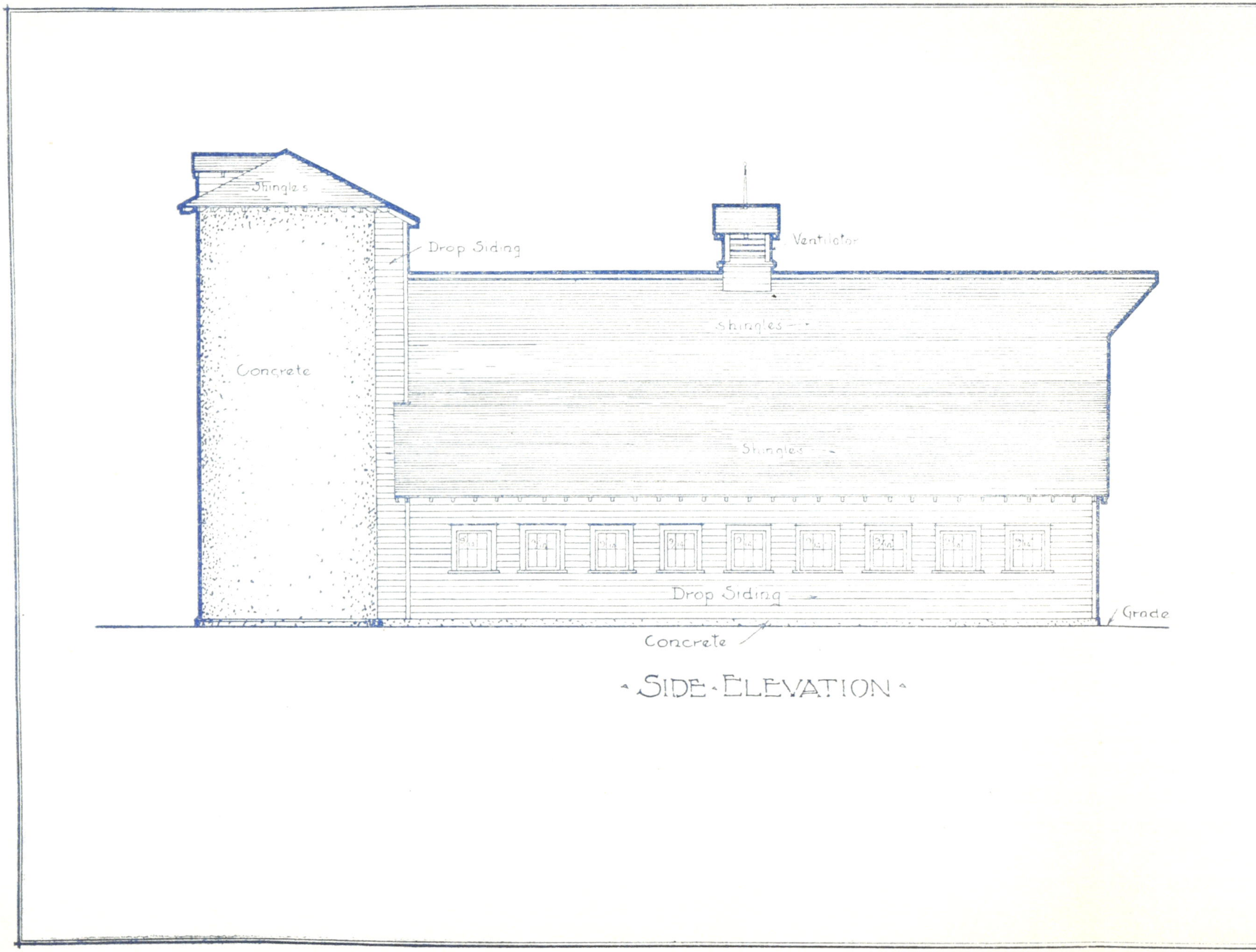

 WHS150111 (one image, split)

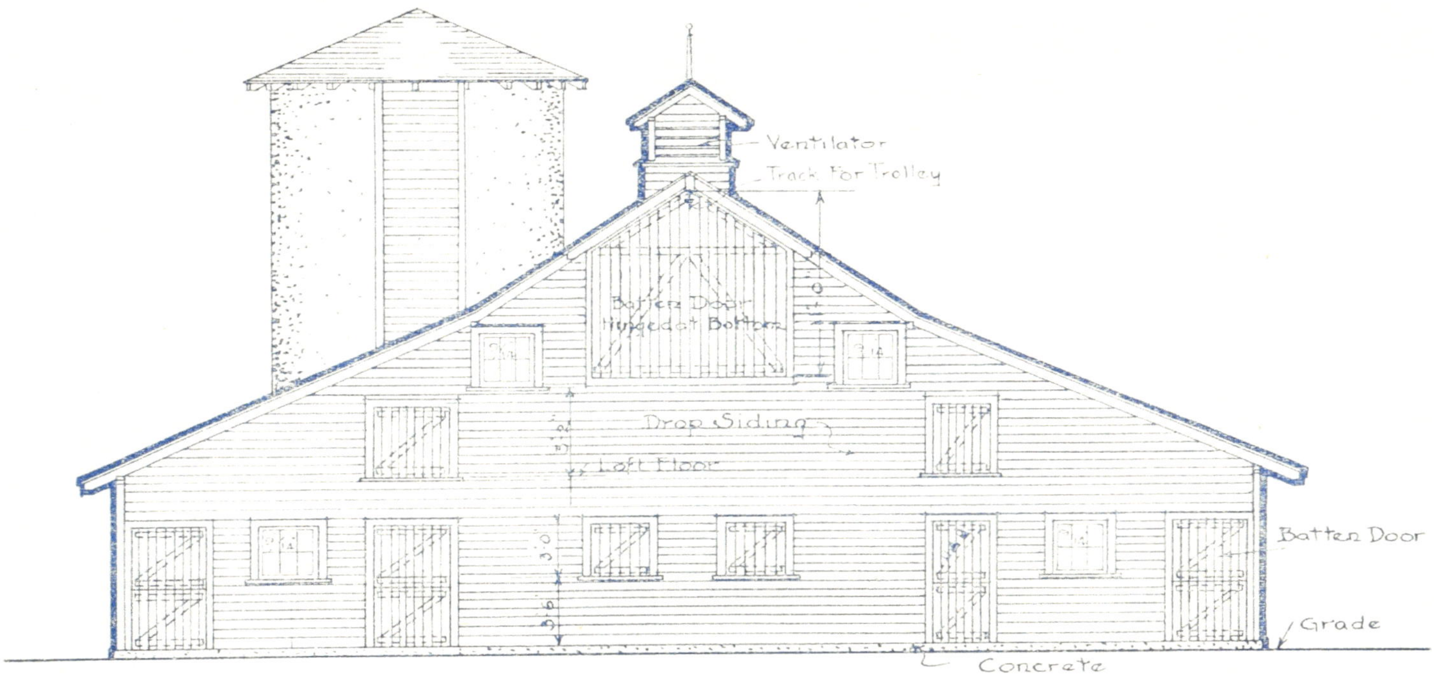

· FRONT · ELEVATION ·

· PLANS · FOR ·
· GENERAL · FARM · BARN ·
· INTERNATIONAL · HARVESTER ·
· COMPANY OF AMERICA ·

SCALE
12" 2' 4' 6' 8' 10'

SHEET No. 1

- REAR - ELEVATION -

 WHS150112 (one image, split)

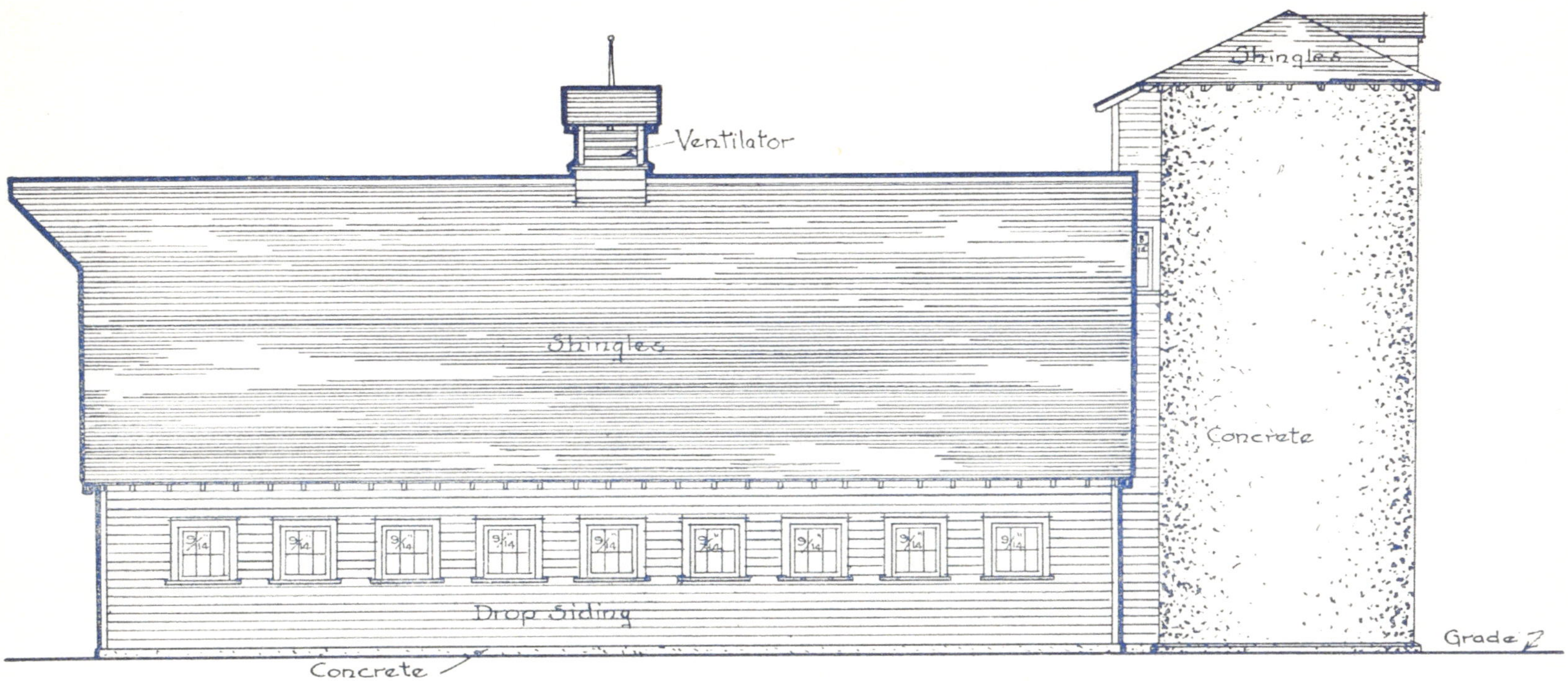

SIDE ELEVATION

PLANS FOR
GENERAL FARM BARN
INTERNATIONAL HARVESTER
COMPANY OF AMERICA

SCALE

SHEET Nº 2

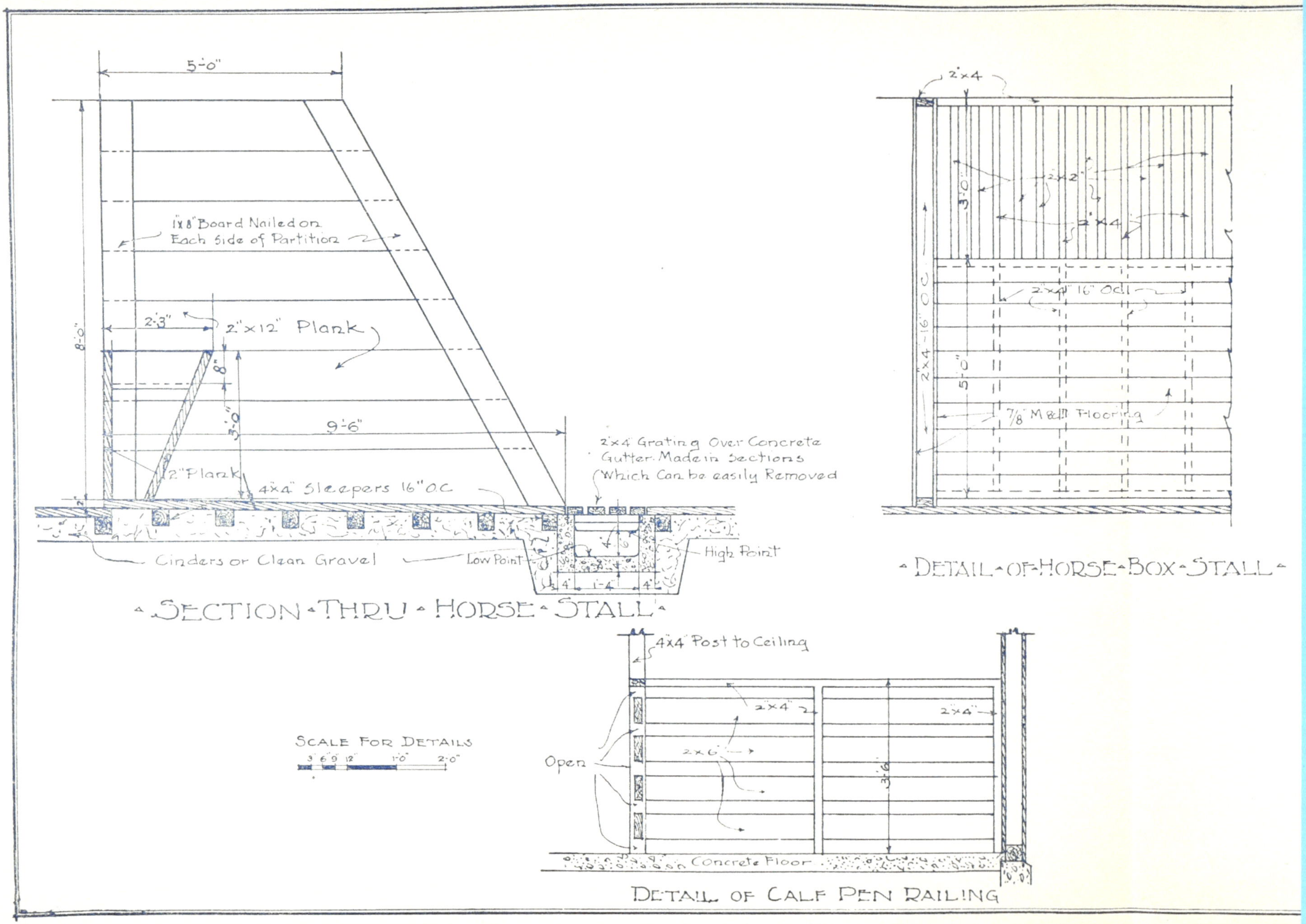
5'-0"
1"x8" Board Nailed on Each Side of Partition
8'-0"
2'-3"
2"x12" Plank
8"
3'-0"
9'-6"
2"x4" Grating Over Concrete Gutter Made in Sections Which Can be easily Removed
2" Plank
4"x4" Sleepers 16" O.C.
Cinders or Clean Gravel
Low Point
High Point
4"
1'-4"
4"
SECTION THRU HORSE STALL
2"x4"
2"x2"
3'-0"
2"x4"
2"x4" 16" O.C.
2"x4" 16" O.C.
5'-0"
7/8" M&B Flooring
DETAIL OF HORSE BOX STALL
4"x4" Post to Ceiling
2"x4"
2"x4"
2"x6"
Open
3'-6"
Concrete Floor
DETAIL OF CALF PEN RAILING
SCALE FOR DETAILS
3" 6" 9" 12" 1'-0" 2'-0"

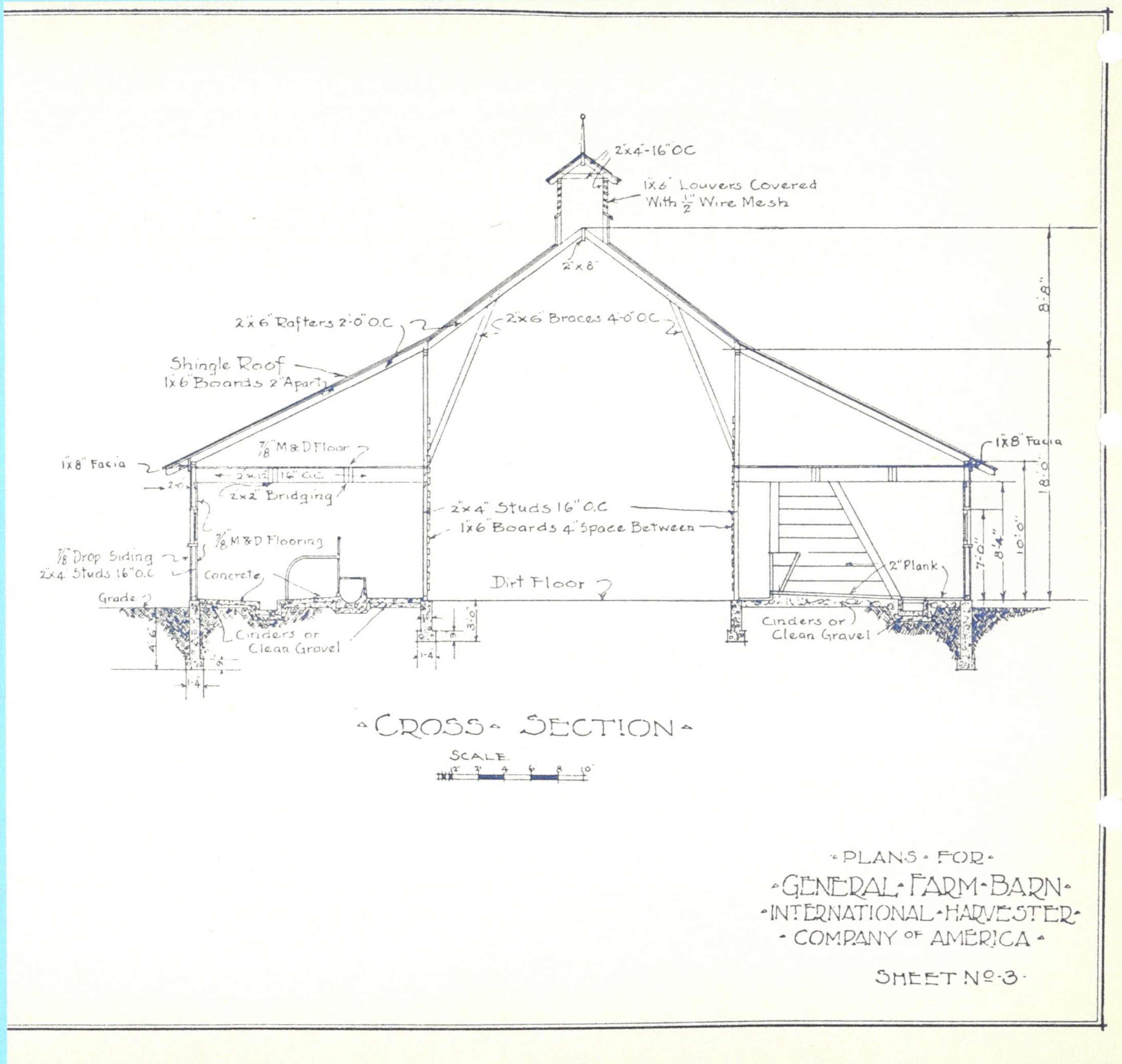

WHS150113 (one image, split)

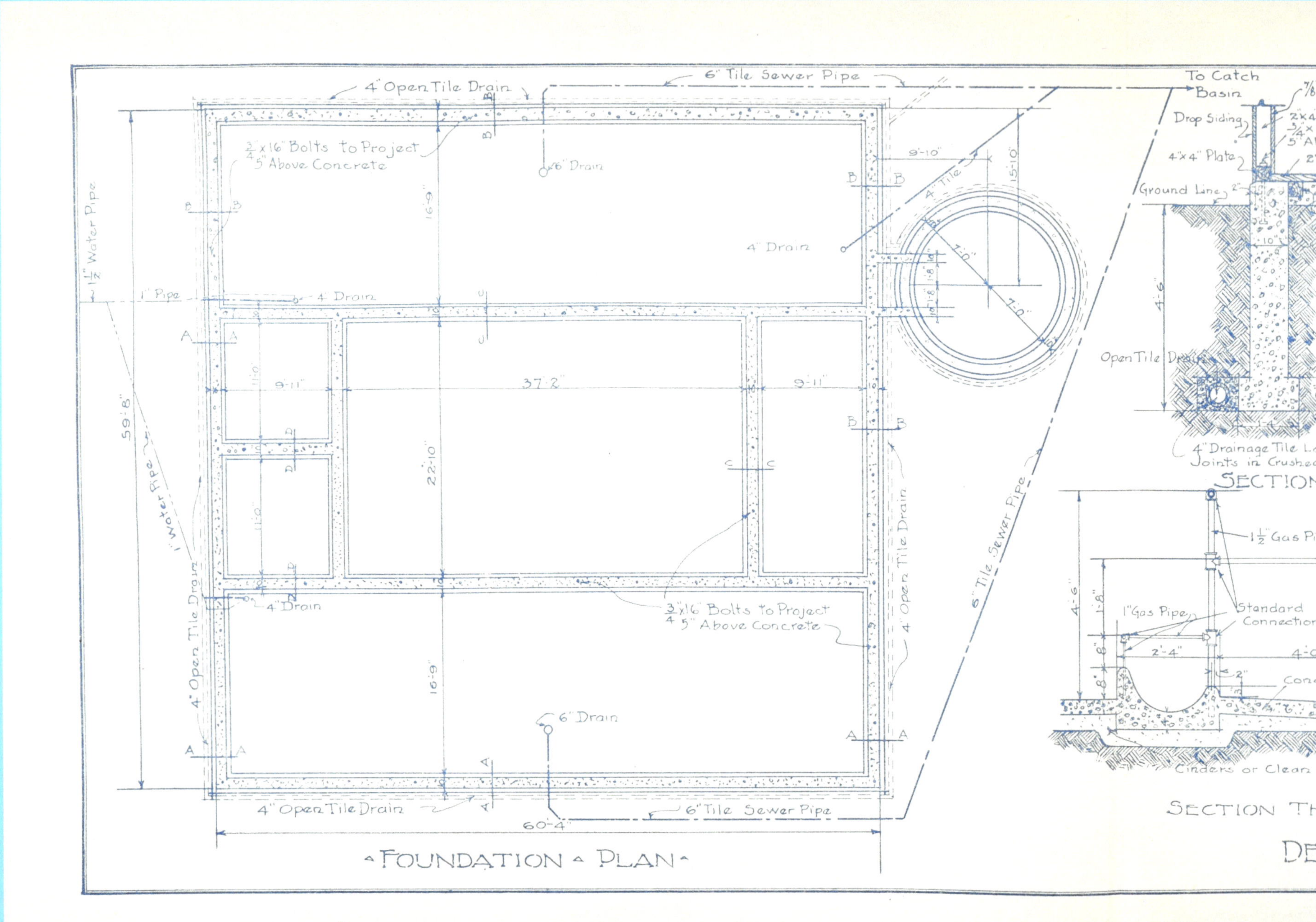

WHS150114 (one image, split)

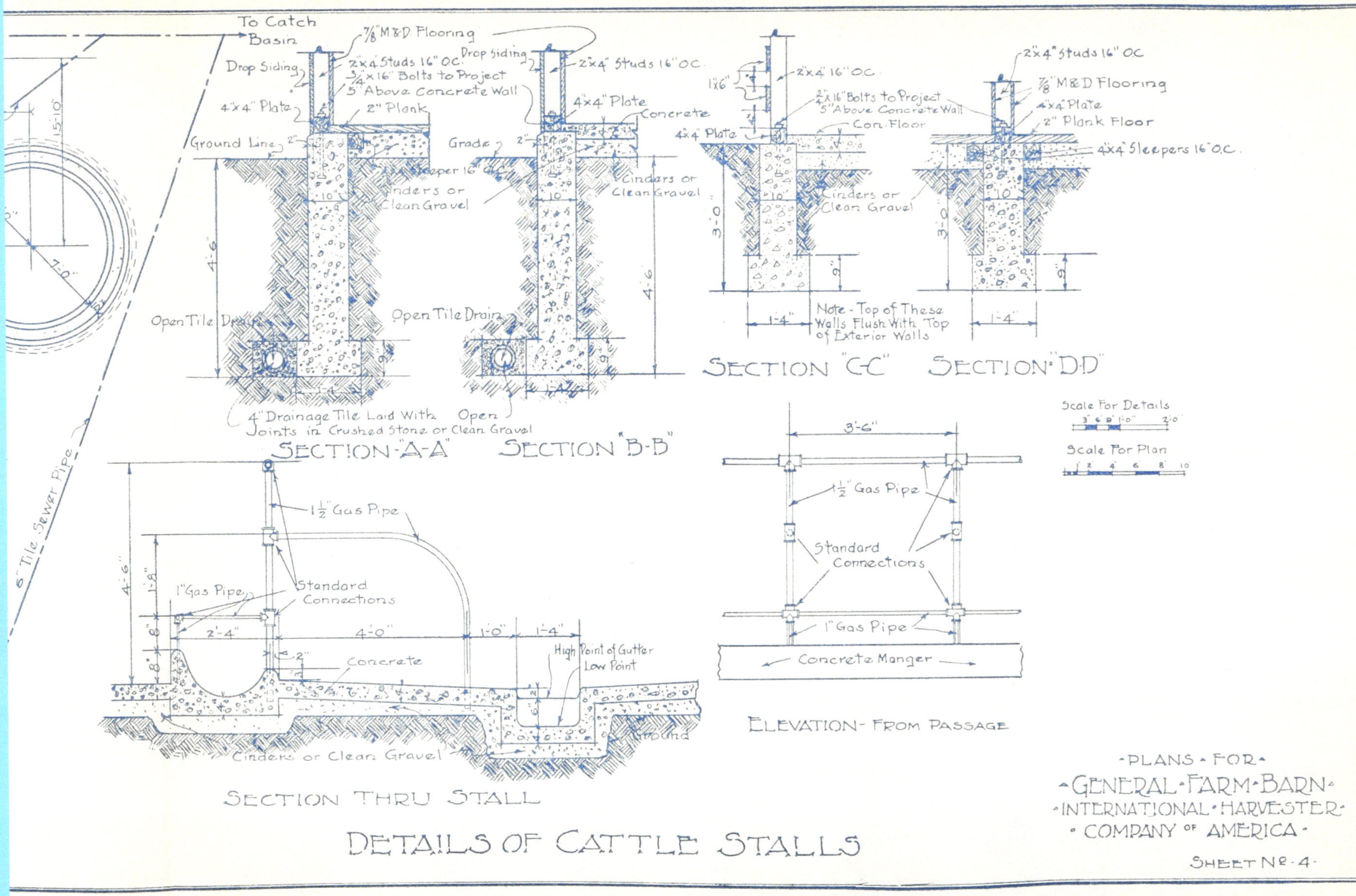
To Catch Basin
7/8" M&D Flooring
Drop Siding
2"x4" Studs 16" O.C.
3/4"x16" Bolts to Project 5" Above Concrete Wall
4"x4" Plate
2" Plank
Ground Line
Concrete
Grade
Cinders or Clean Gravel
Open Tile Drain
4" Drainage Tile Laid With Open Joints in Crushed Stone or Clean Gravel
SECTION "A-A"
SECTION "B-B"
2"x4" 16" O.C.
1"x6"
Con. Floor
Note - Top of These Walls Flush With Top of Exterior Walls
2" Plank Floor
4"x4" Sleepers 16" O.C.
SECTION "C-C"
SECTION "D-D"
6" Tile Sewer Pipe
1 1/2" Gas Pipe
1" Gas Pipe
Standard Connections
High Point of Gutter
Low Point
Ground
Cinders or Clean Gravel
SECTION THRU STALL
Concrete Manger
ELEVATION - FROM PASSAGE
Scale For Details
Scale For Plan
DETAILS OF CATTLE STALLS
PLANS FOR
GENERAL FARM BARN
INTERNATIONAL HARVESTER COMPANY OF AMERICA
SHEET No. 4

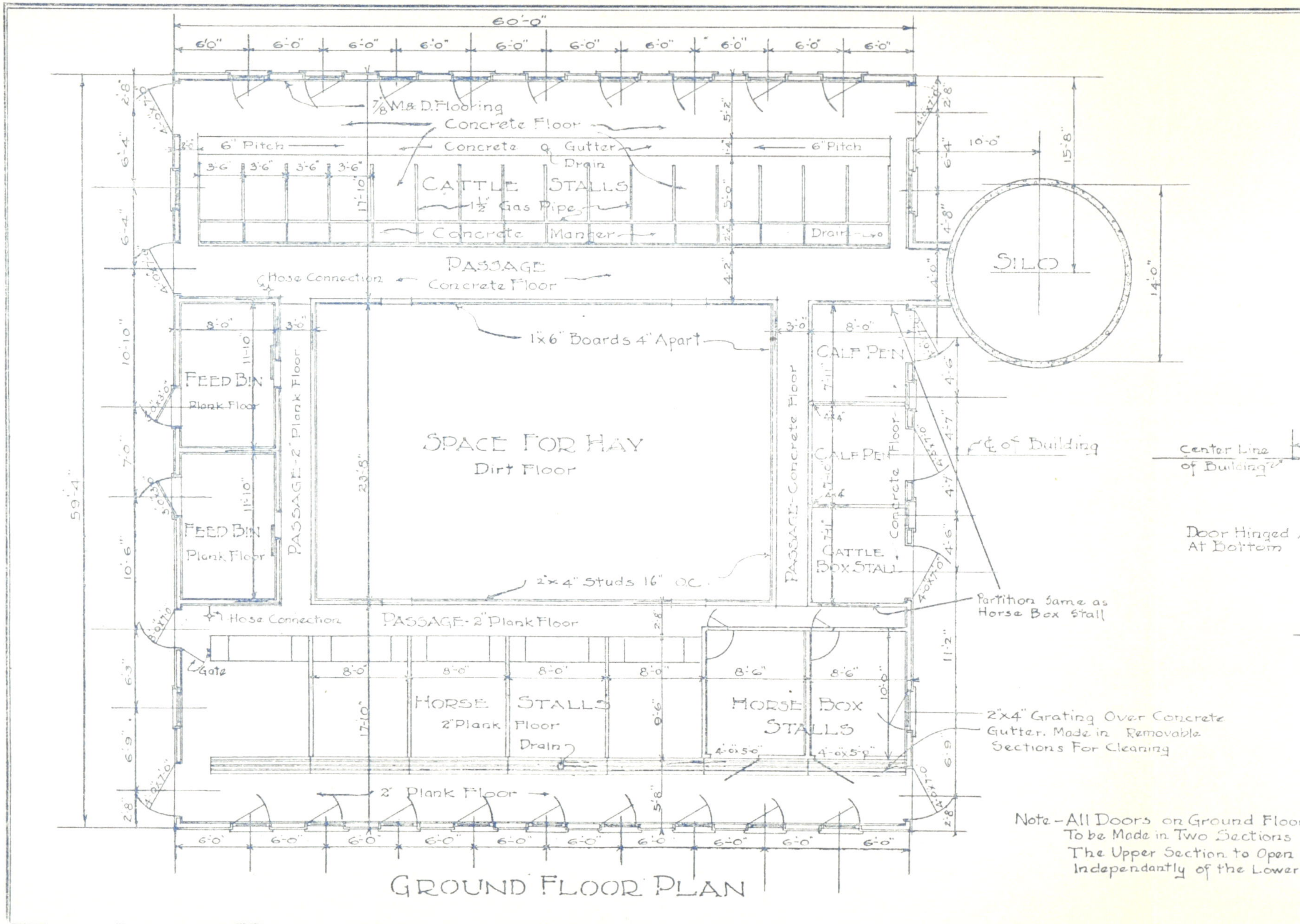

GROUND FLOOR PLAN

 WHS150115 (one image, split)

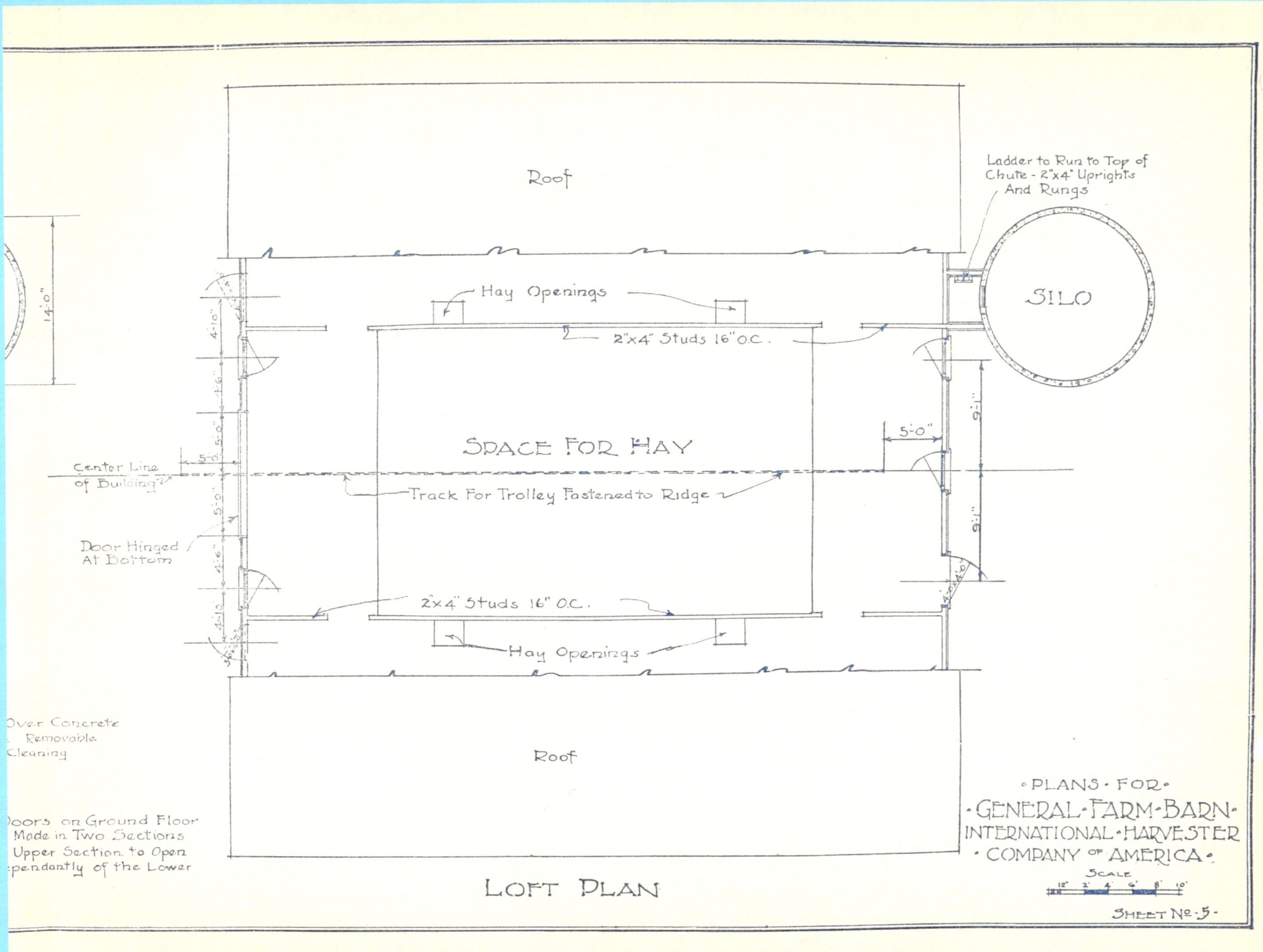
Roof
Ladder to Run to Top of Chute - 2"x4" Uprights And Rungs
Hay Openings
SILO
2"x4" Studs 16" O.C.
14'-0"
4'-10"
4'-6"
5'-0"
5'-0"
9'-1"
Center Line of Building
SPACE FOR HAY
Track For Trolley Fastened to Ridge
Door Hinged At Bottom
2"x4" Studs 16" O.C.
Hay Openings
Over Concrete
Removable
Cleaning
Roof
Doors on Ground Floor Made in Two Sections Upper Section to Open
pendantly of the Lower
LOFT PLAN
PLANS FOR
GENERAL FARM BARN
INTERNATIONAL HARVESTER
COMPANY OF AMERICA
SCALE
SHEET No. 5

WHS150116 (one image, split)

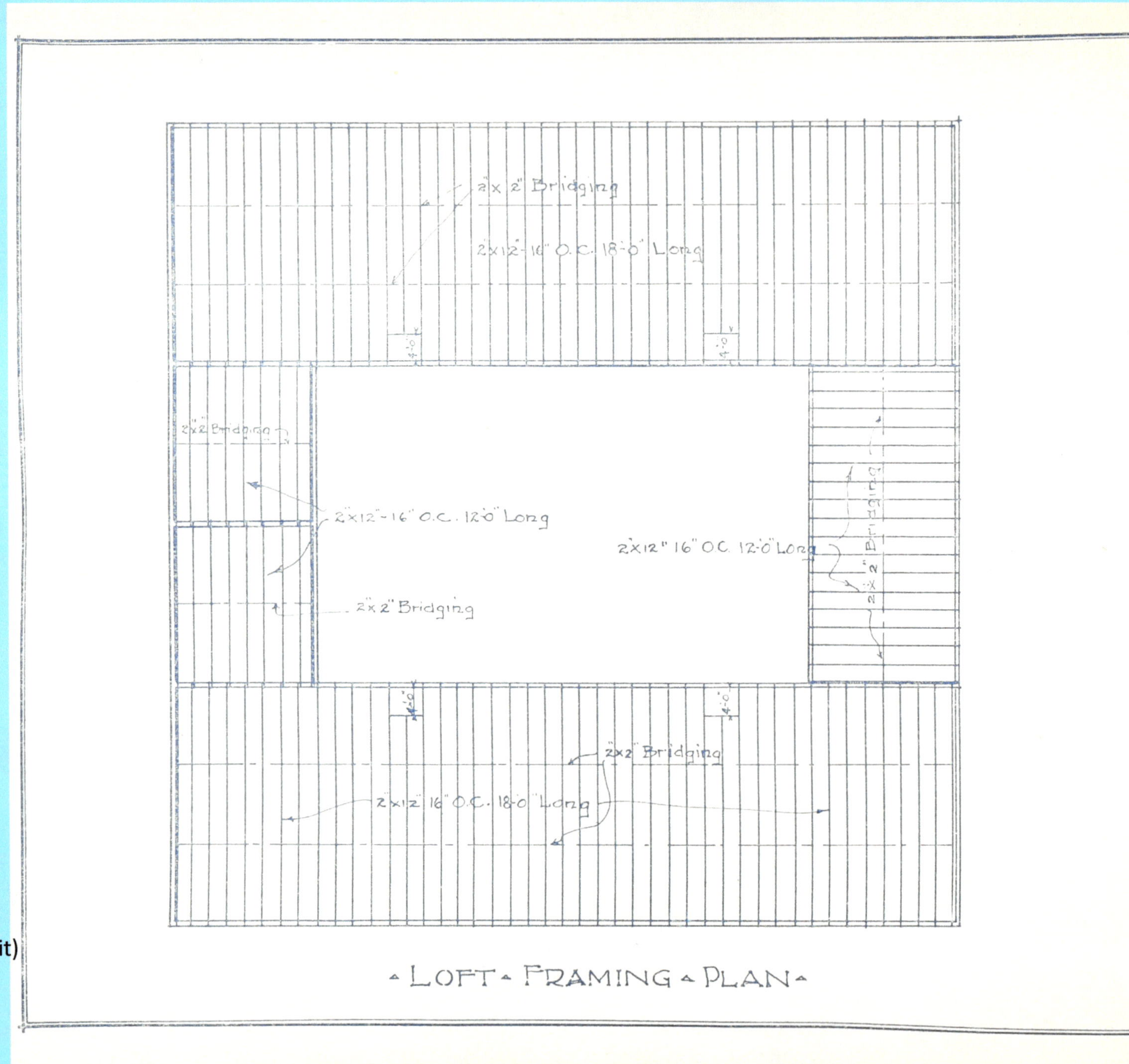

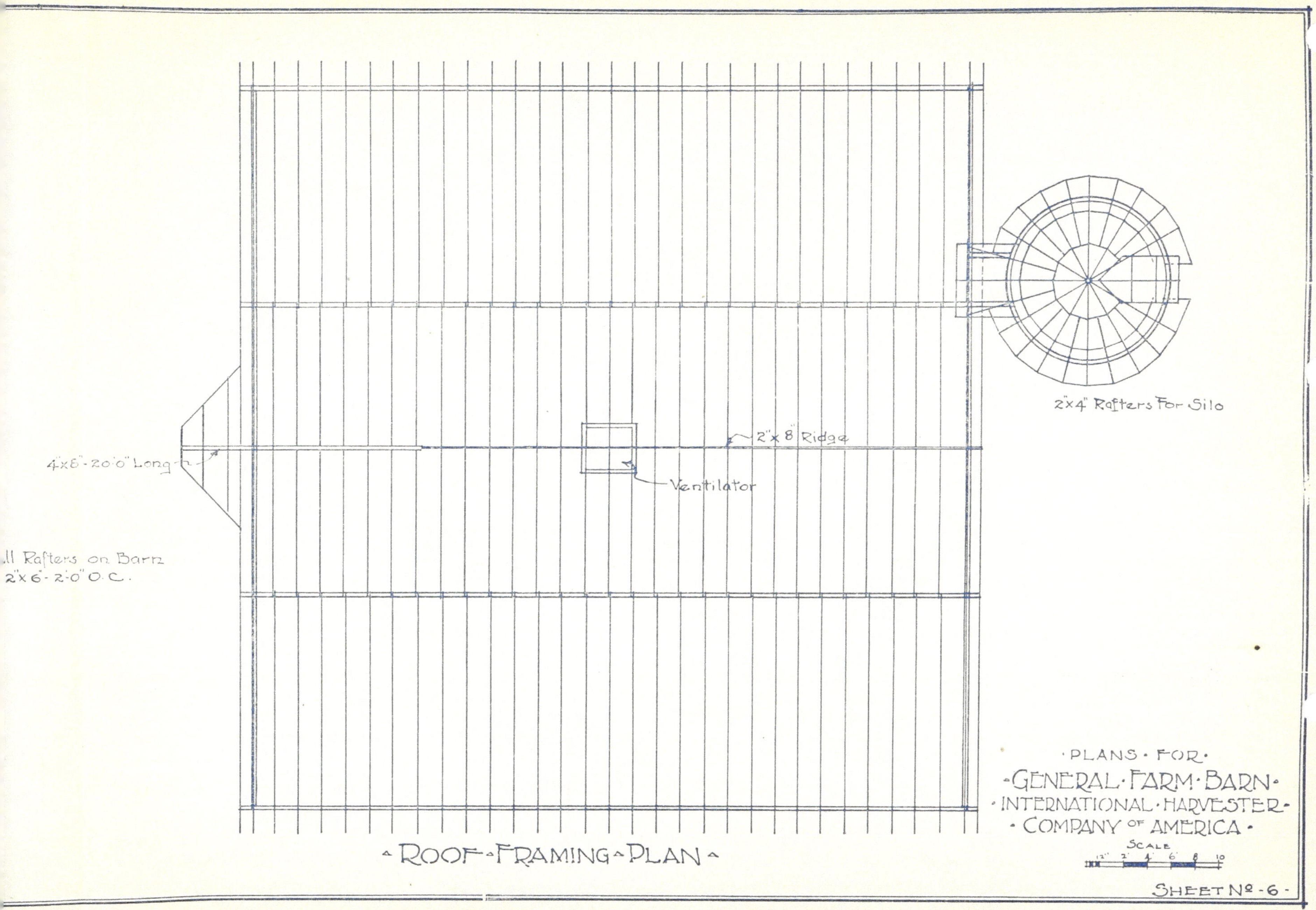
2"x4" Rafters For Silo
2"x 8" Ridge
4"x8"-20'0" Long
Ventilator
ll Rafters on Barn
2"x6-2'0" O.C.
· PLANS · FOR ·
· GENERAL · FARM · BARN ·
· INTERNATIONAL · HARVESTER ·
· COMPANY OF AMERICA ·
SCALE
SHEET Nº - 6 -
ROOF · FRAMING · PLAN

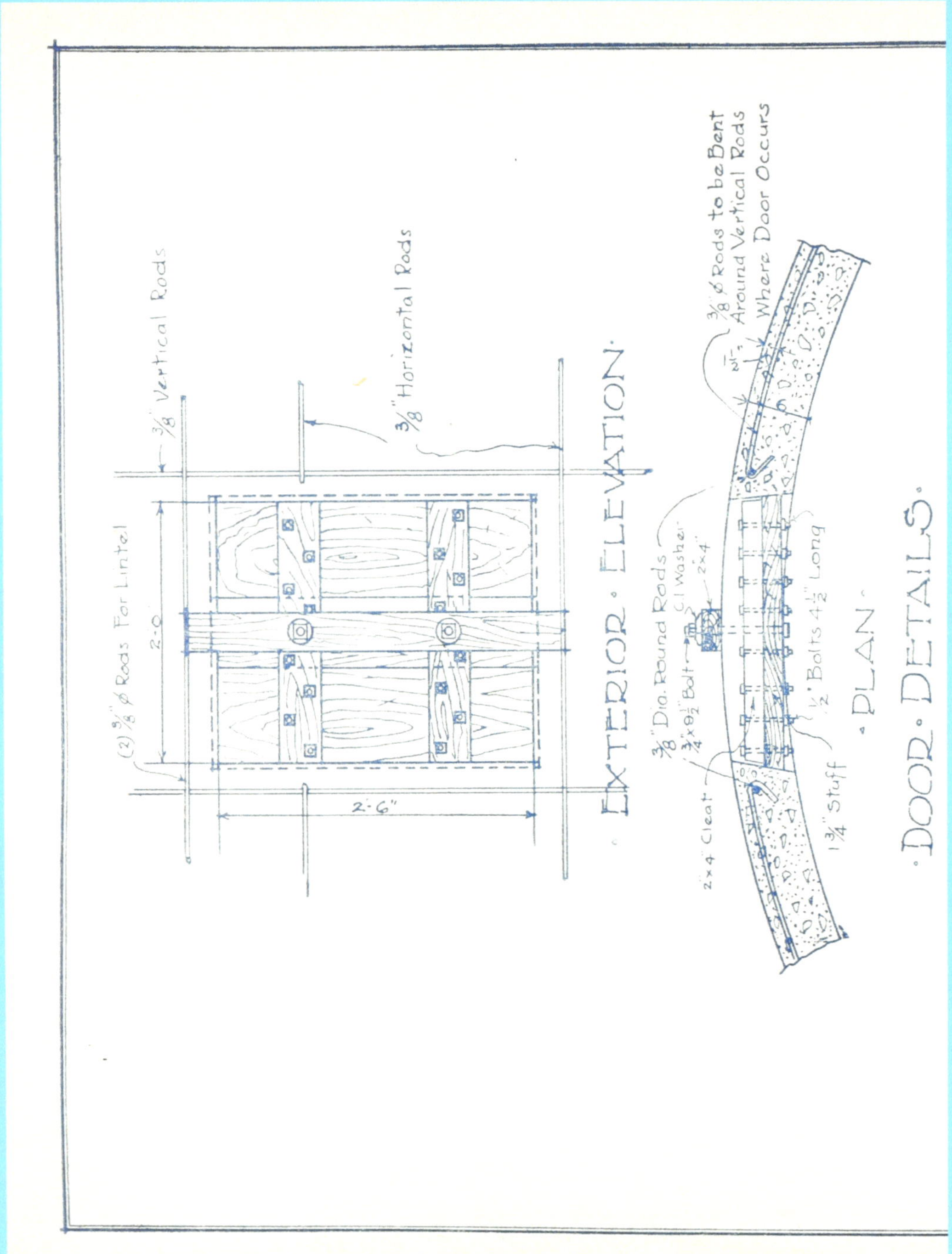

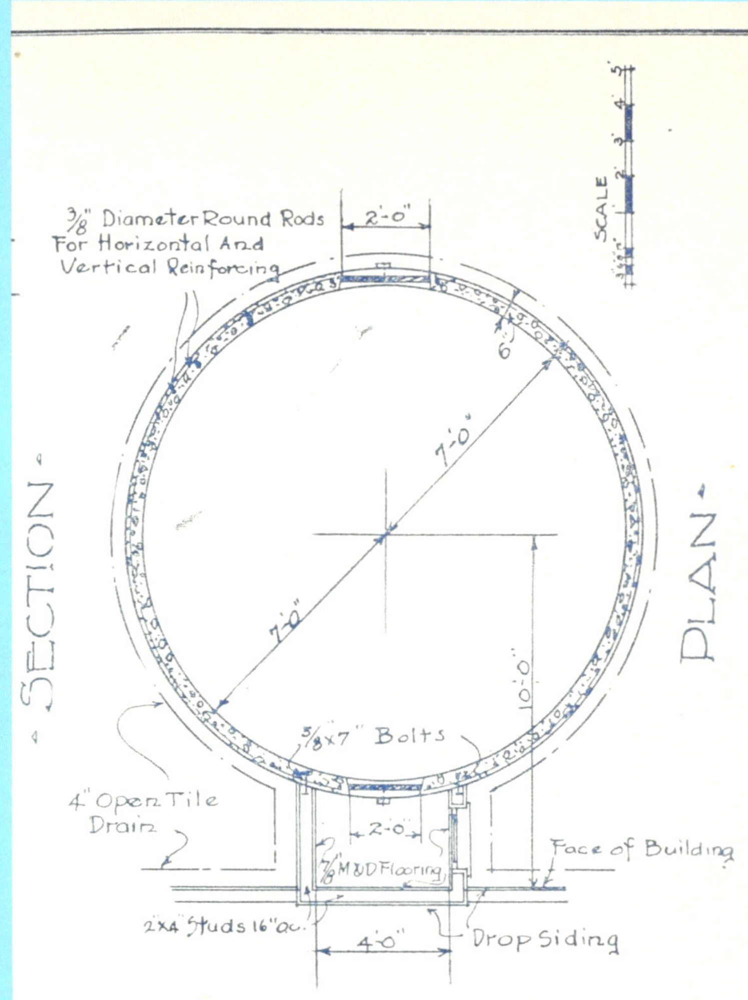

The silo shown here is one version of the many different options that International Harvester promoted.

WHS150117 (one image, both pages, split)

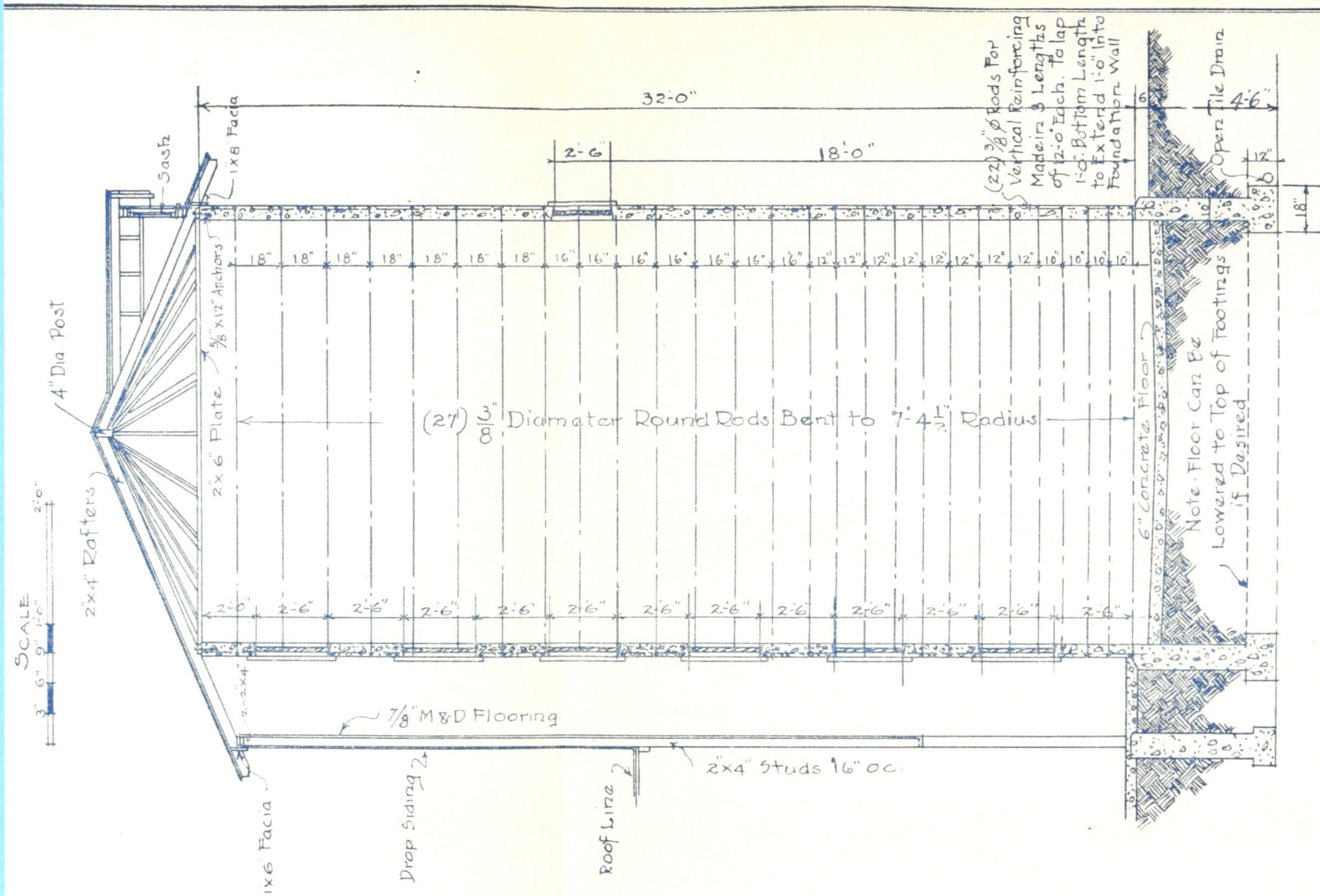

·DETAILS · OF· REINFORCED· CONCRETE · SILO·

·PLANS · FOR·
·GENERAL·FARM·BARN·
·INTERNATIONAL·HARVESTER·
·COMPANY OF AMERICA·

SHEET Nº 7·

Depending on the geographical area, International Harvester recommended wood, cement, tile, or pit styles of silos. Some of the silo designs were referred to by the location they were most commonly used, or the materials they were built from.
Silos were all various sizes, depending on the amount of feed needed to store, which Harvester also helped teach how to calculate.

WHS147499

BUILDING PLANS

AND

SPECIFICATIONS

FOR BUILDING

SILOS

ISSUED AND COPYRIGHTED 1912 BY

INTERNATIONAL HARVESTER COMPANY OF AMERICA
(INCORPORATED)

CHICAGO USA

STAVE SILO

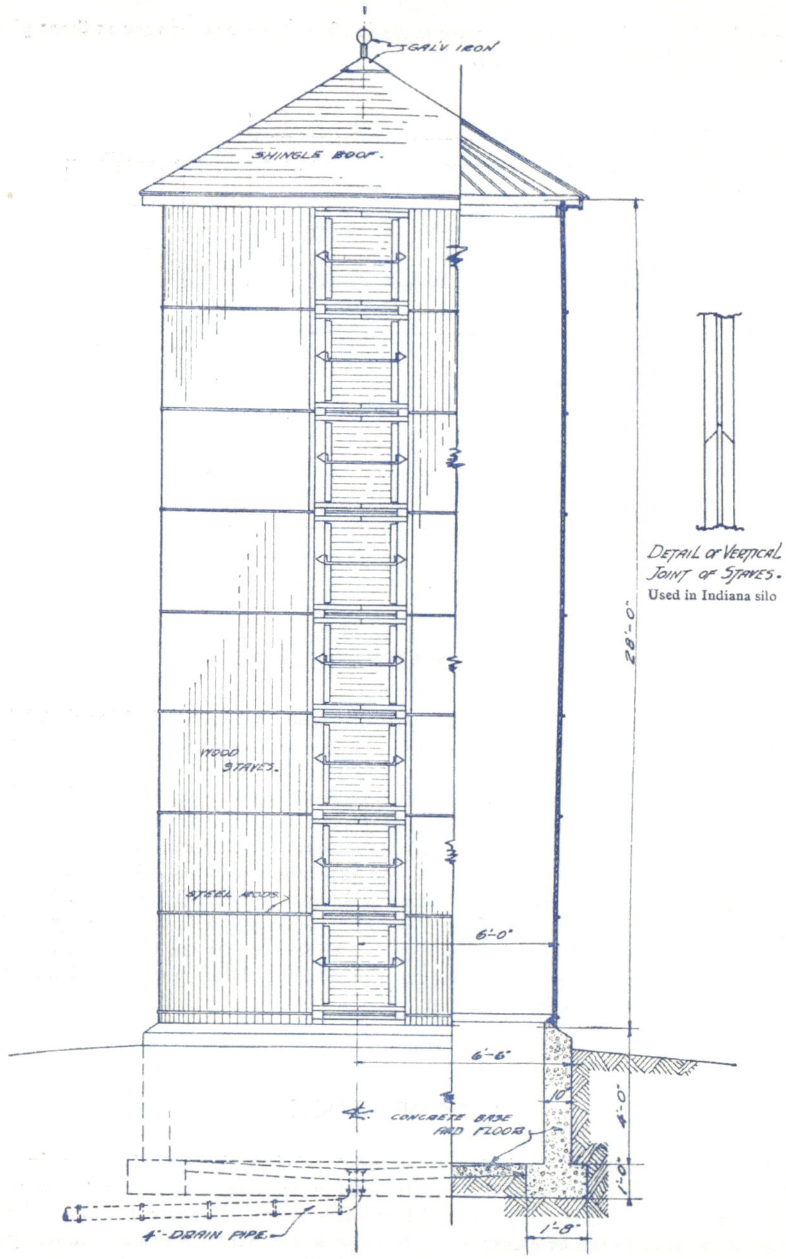

Fig. 20. Section stave silo

WHS150150 (both pages, split, one image)

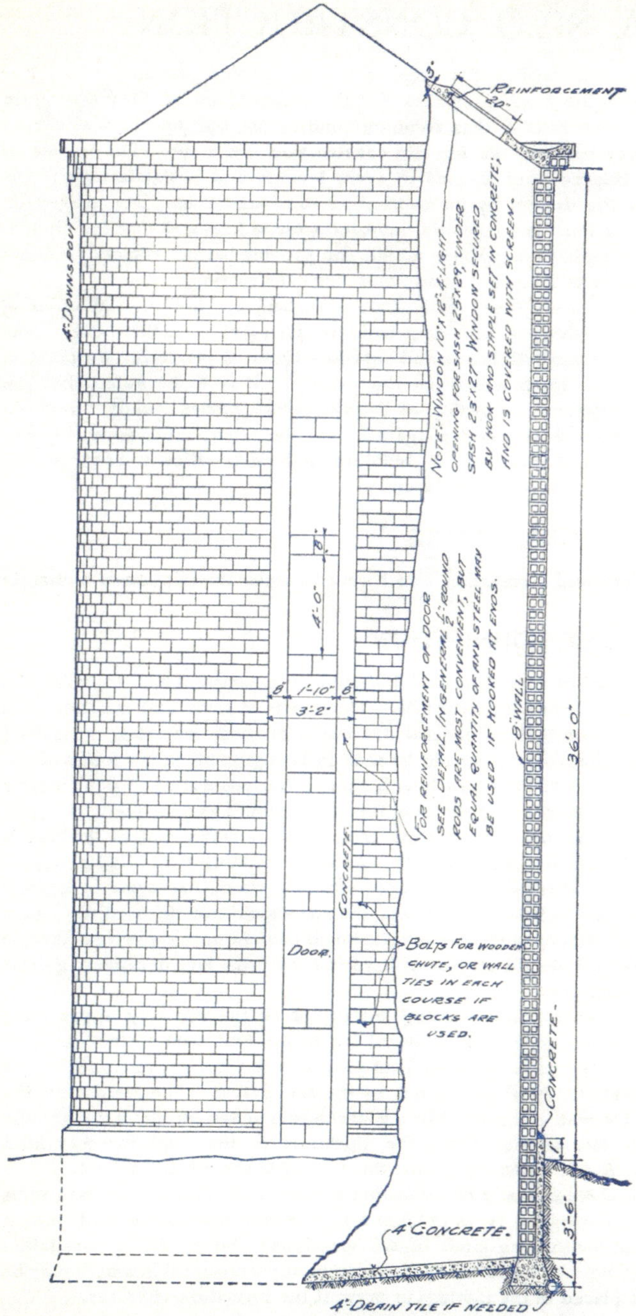

Fig. 1. Section of Iowa silo

IOWA SILO

(HOLLOW TILE TYPE)

Mixture to Use for Foundations

1 : 2½ : 5

For Mortar

1 : 3

For Roof and Door Frames

1 : 2½ : 4 Concrete

For Washing Inside

1 : 1

CONCRETE BLOCK SILO DETAILS

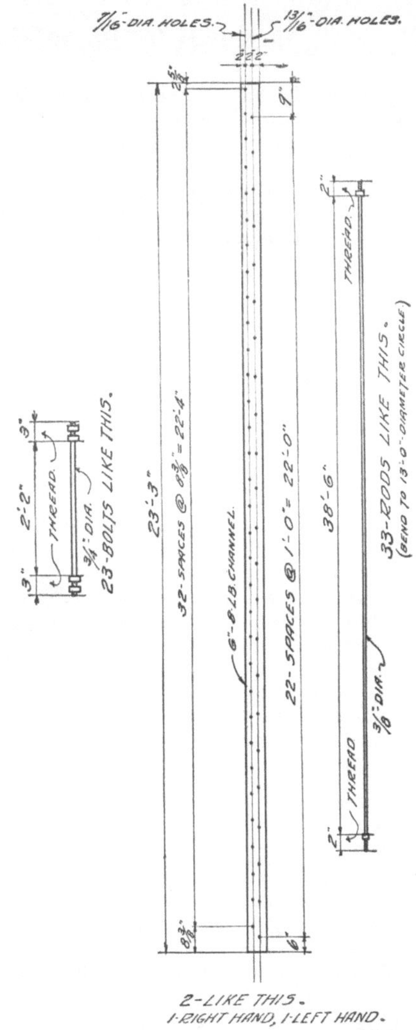

Fig. 15. Reinforcing detail for concrete silo

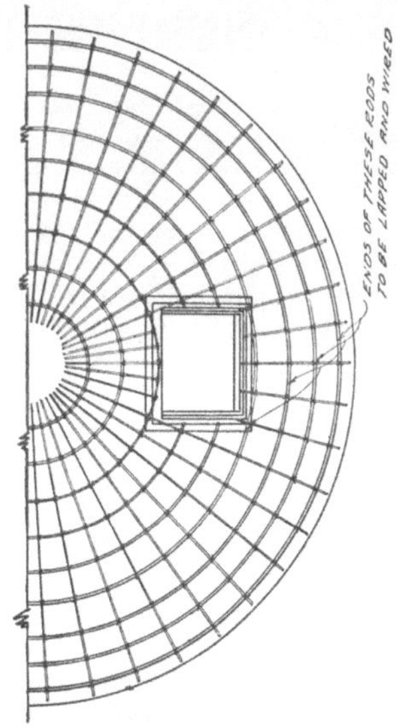

Fig. 17. Plan of roof showing location of reinforcing rods

LIST OF RODS FOR ROOF.

50-1/4" DIA. -	7'-4" -	LG.
2-3/8" "	23'-0"	"
1-1/4" "	18'-0"	"
1-1/4" "	16'-0"	"
1- " "	14'-0"	"
1- " "	12'-0"	"
1- " "	10'-0"	"
1- " "	7'-0"	"
1- " "	5'-0"	"

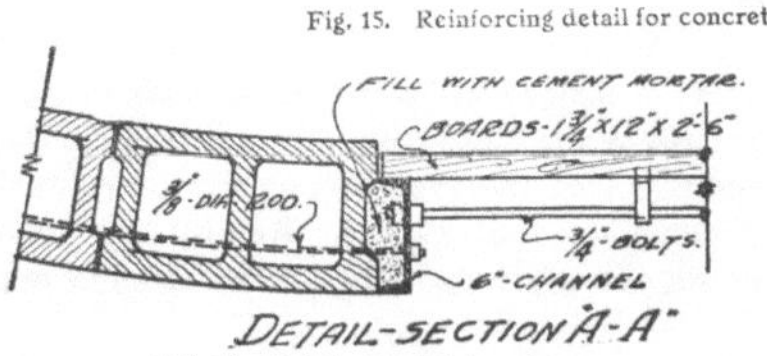

Fig. 16. Details showing door frame

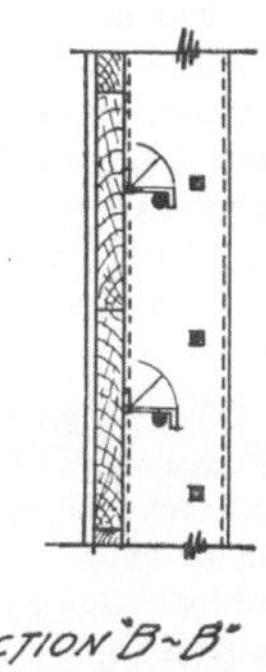

Fig. 18. Details section "B-B" showing door

IOWA SILO DETAILS

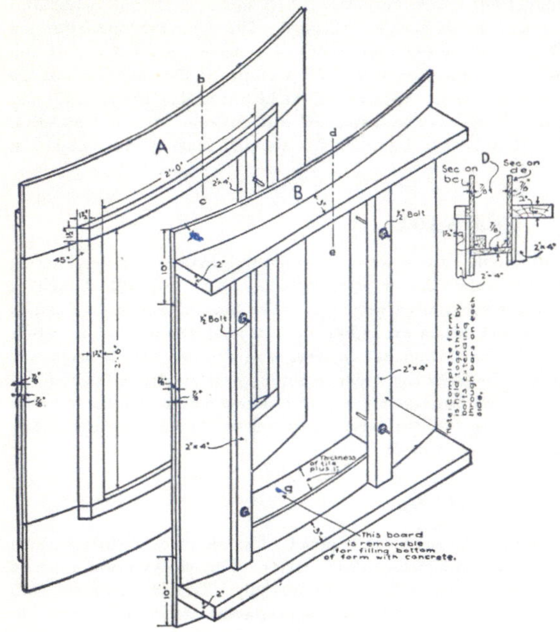

Fig. 2. Forms of concrete door frame for Iowa silo

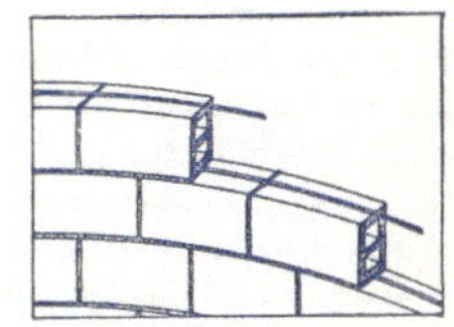

Fig. 5. Showing location of reinforcing rods

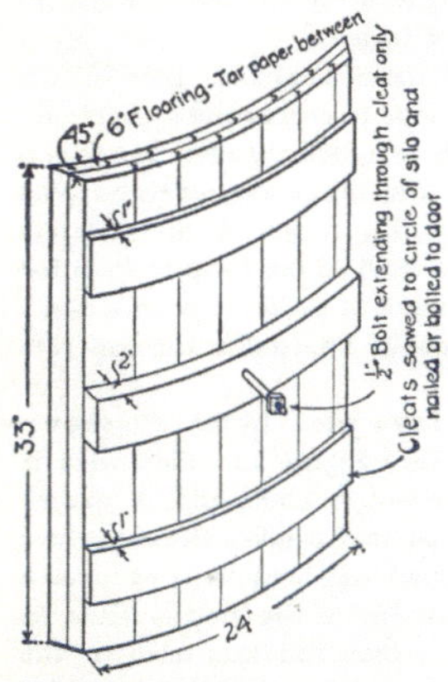

Fig. 3. Door for Iowa or concrete silo

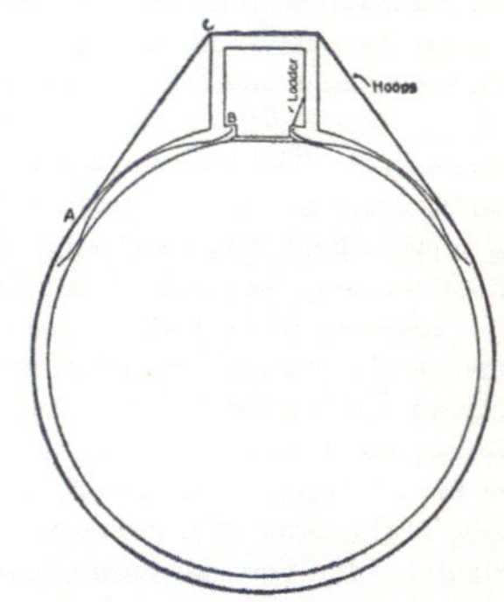

Fig. 4. Construction showing continuous door for Iowa silo

CONCRETE BLOCK SILO

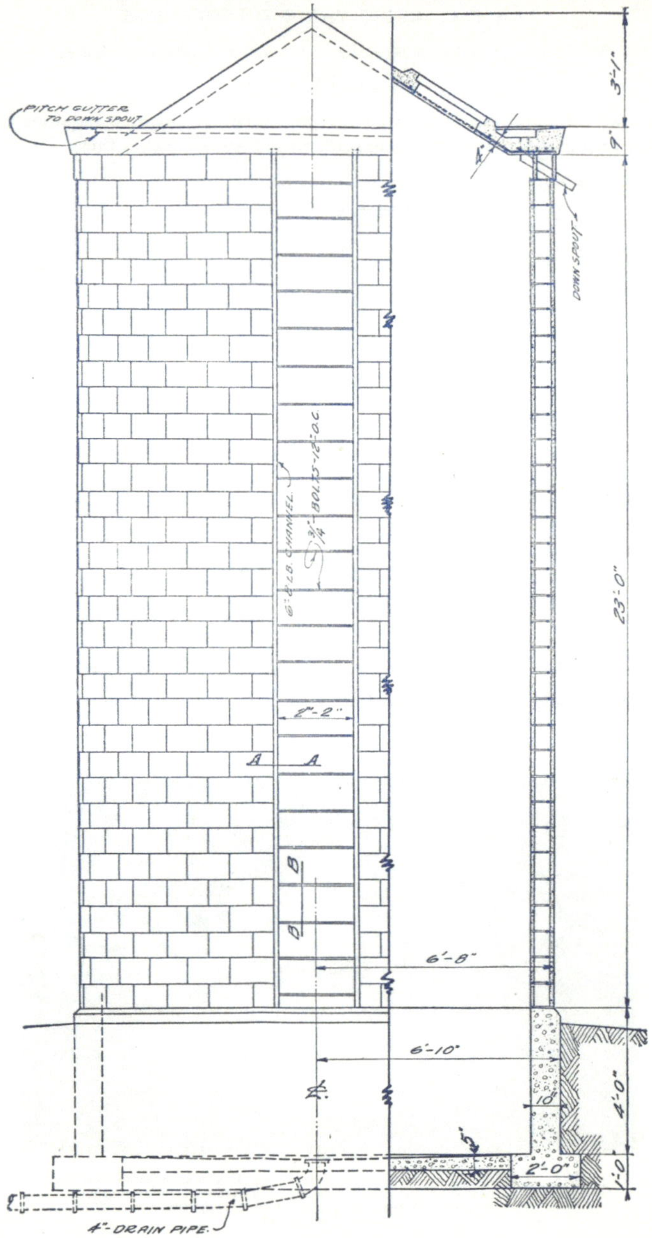

Fig. 14. Section of concrete block silo

CONCRETE SILO DETAILS

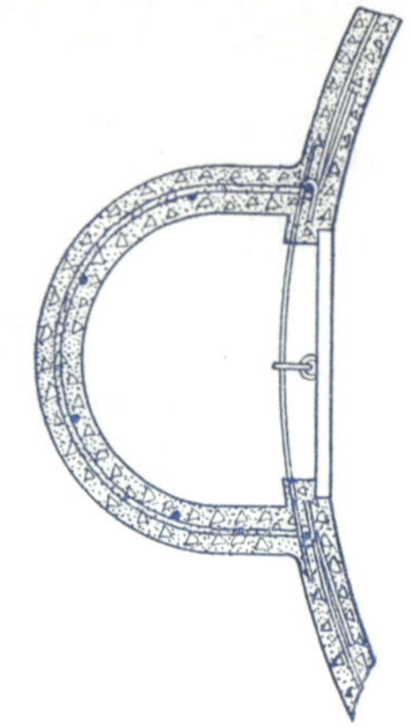

Fig. 7. Section through chute

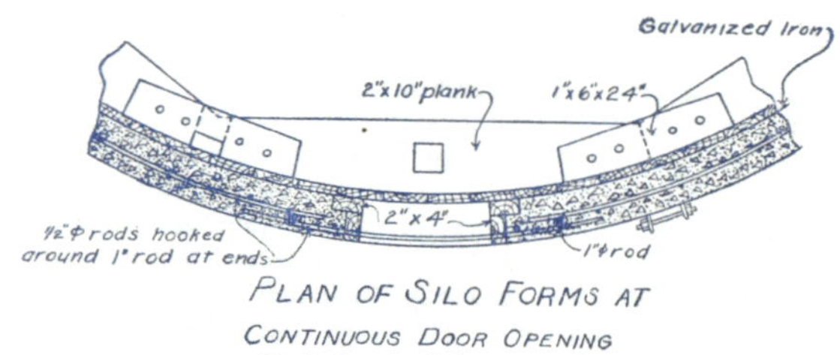

Fig. 8. Door details for concrete silo

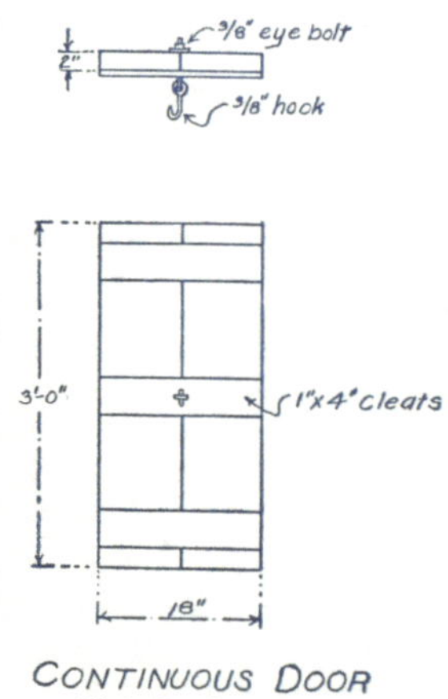

Fig. 9. Door for concrete silo

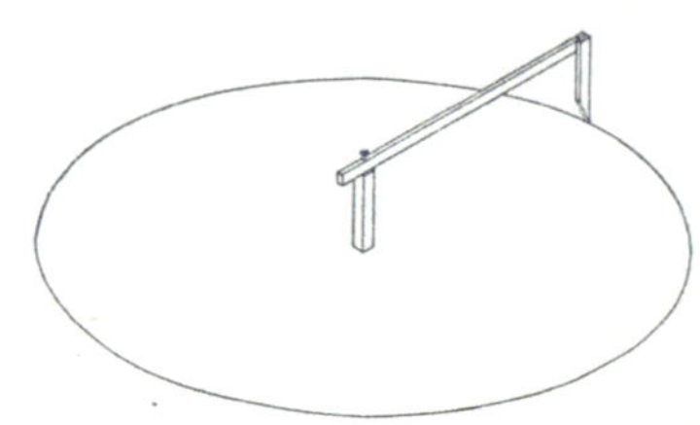

Fig. 10. Simple sweep, convenient in laying out excavation

CONCRETE SILO

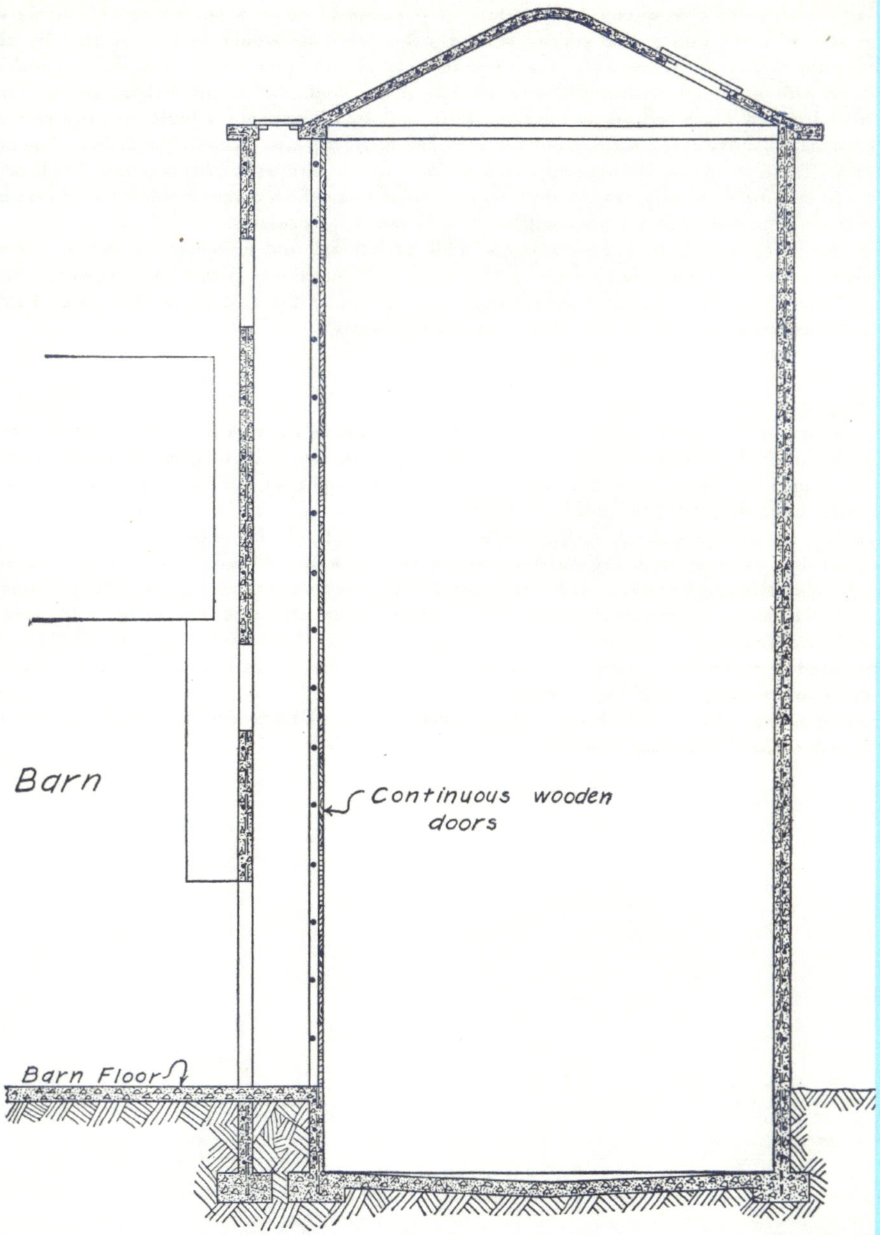

Fig. 6. Section of concrete silo

DETAILS OF CONCRETE SILO FORMS

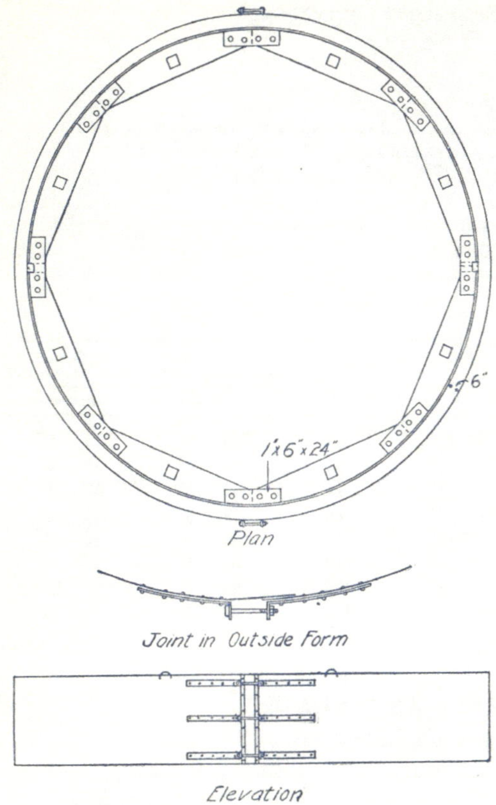

Fig. 11. Plan of home made wall forms, and details of outside form

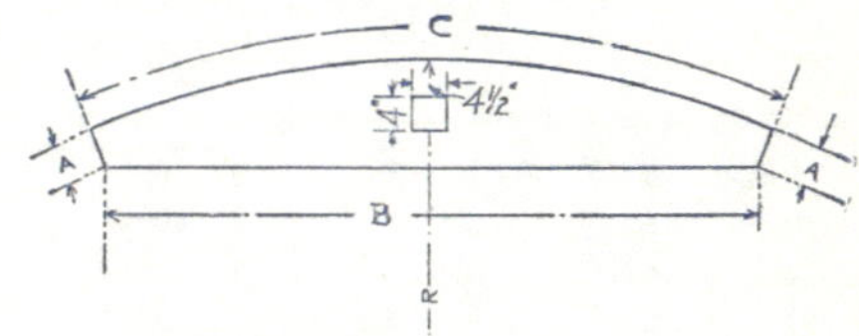

Fig. 13. Inner form ribs

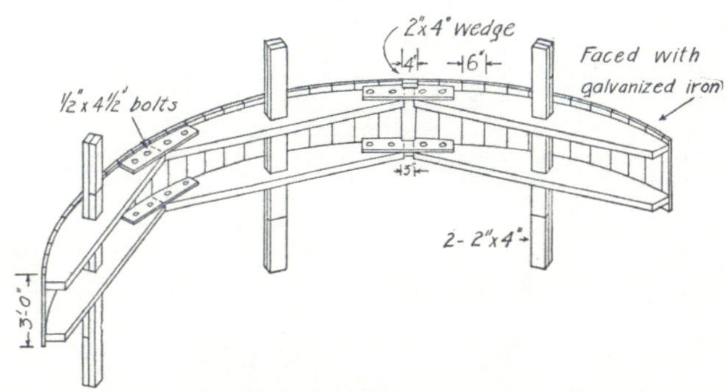

Fig. 12. Perspective of inner wall form, showing position of 2 in. x 4 in. uprights upon which the form is raised

International Harvester in other countries also provided plans for anyone that requested. This cow shed or daiy plan has four different views of a simple shed for milking cows in Australia.
Tomac Collection

COW SHED PLAN

ABBREVIATIONS—

Boiler	B.
Engine	E.
Pulsator Shafting	P.S.
Separator	Sep.
Troughs	T.
Vacuum Pump	V.P.
Vacuum Tank	V.T.
Rack	R.
Counter Shaft	C.

SIZES OF SEPARATOR ROOMS

UP TO 40 COWS . . 10′ x 8′ x 8′
40 TO 80 COWS . . 12′ x 10′ x 8′
OVER 80 COWS . . 12′ x 12′ x 8′

FIXED WINDOWS IN SEPARATOR-ROOM, IN WALLS BETWEEN THE ENGINE AND WASH ROOM, ARE NOT COMPULSORY BUT ADD GREATLY TO CONVENIENCE.

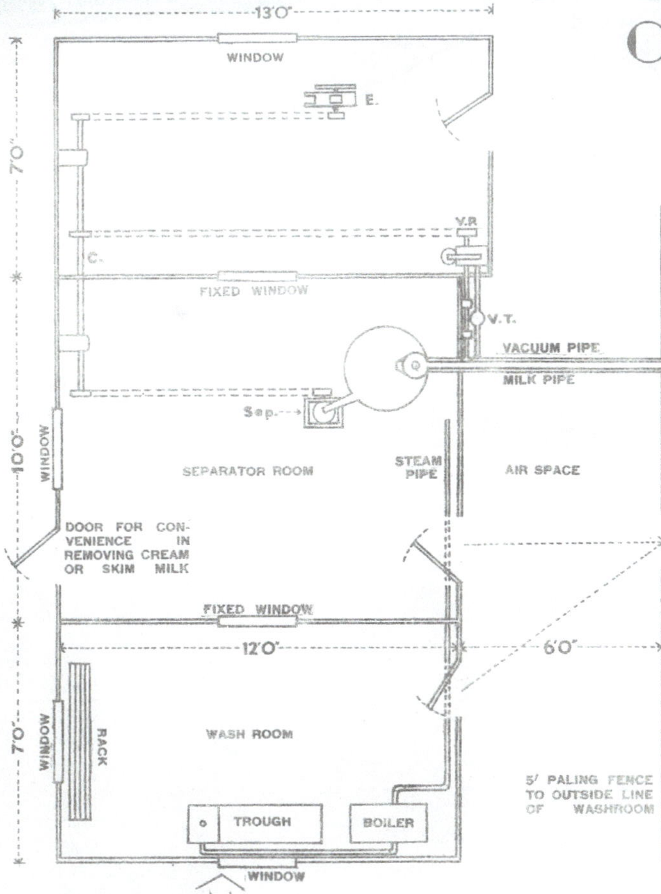

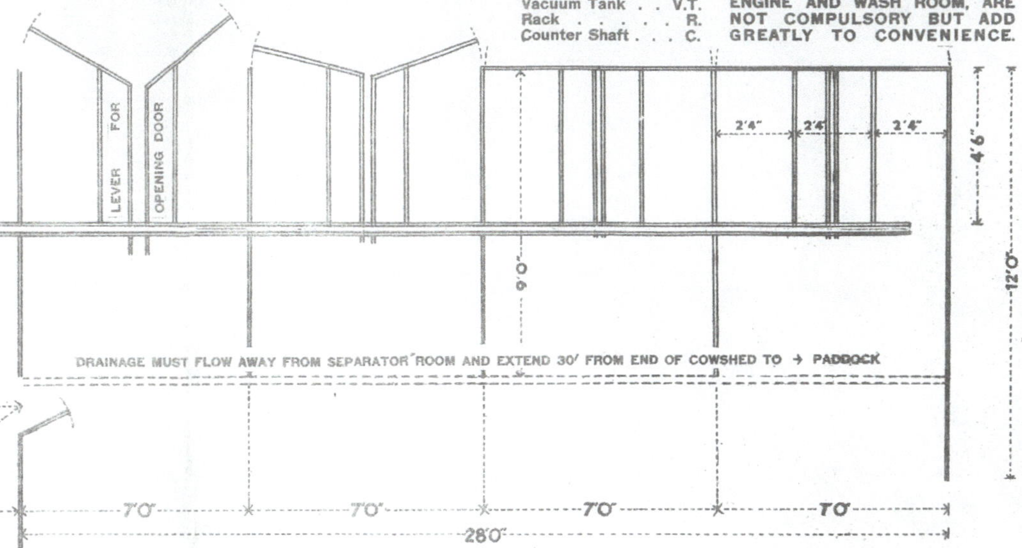

PLAN OF COWSHED FOR MILKING MACHINE COMPLYING WITH REQUIREMENTS OF DEPARTMENT OF AGRICULTURE.

MODIFICATIONS TO SUIT INDIVIDUAL REQUIREMENTS CAN BE MADE BY COLLABORATION WITH LOCAL DAIRY SUPERVISOR.

INTERNATIONAL HARVESTER COMPANY OF AUSTRALIA PTY. LTD.
(INCORPORATED IN VICTORIA)
171-205 CITY ROAD, SOUTH MELBOURNE, S.C.4
ROBT. NETTLEFOLD PTY. LTD., HOBART. AGENTS FOR TAS.

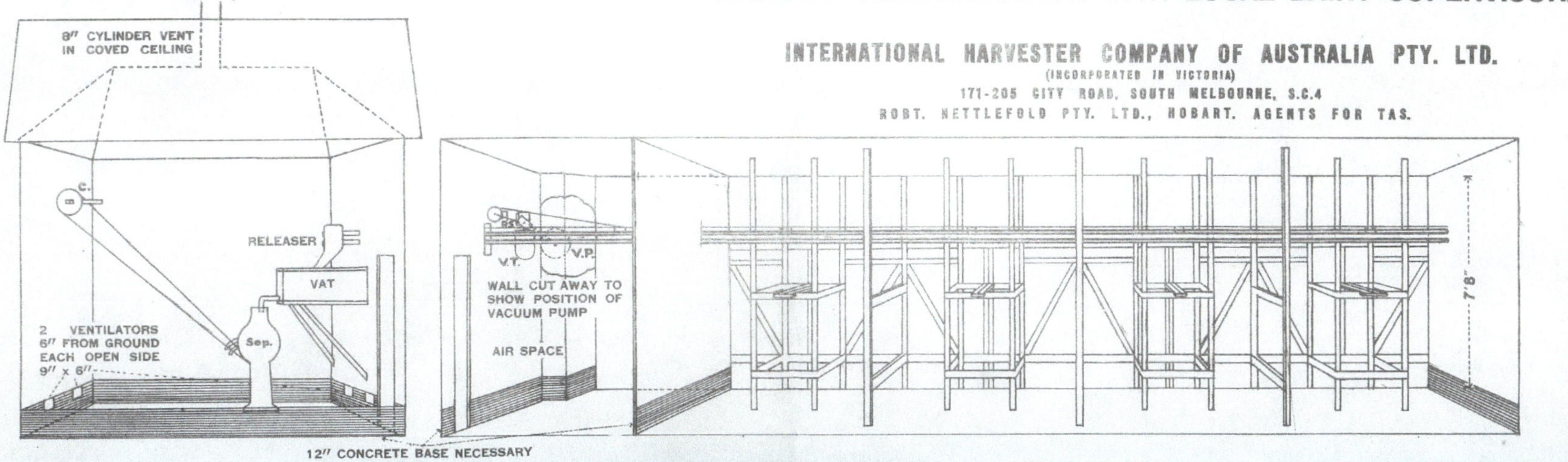

Equipment storage sizes in 1912 were much smaller than modern day requirements. A farmer could fit everything into a small space, not much larger than a two car garage by today standards. The overall dimensions for this building are 24’ by 34’.

WHS147497

BUILDING PLANS

SPECIFICATIONS

AND

MATERIAL LIST

FOR A

MACHINE SHED

ISSUED AND COPYRIGHTED 1913 BY

INTERNATIONAL HARVESTER COMPANY OF AMERICA

(INCORPORATED)

CHICAGO USA

566

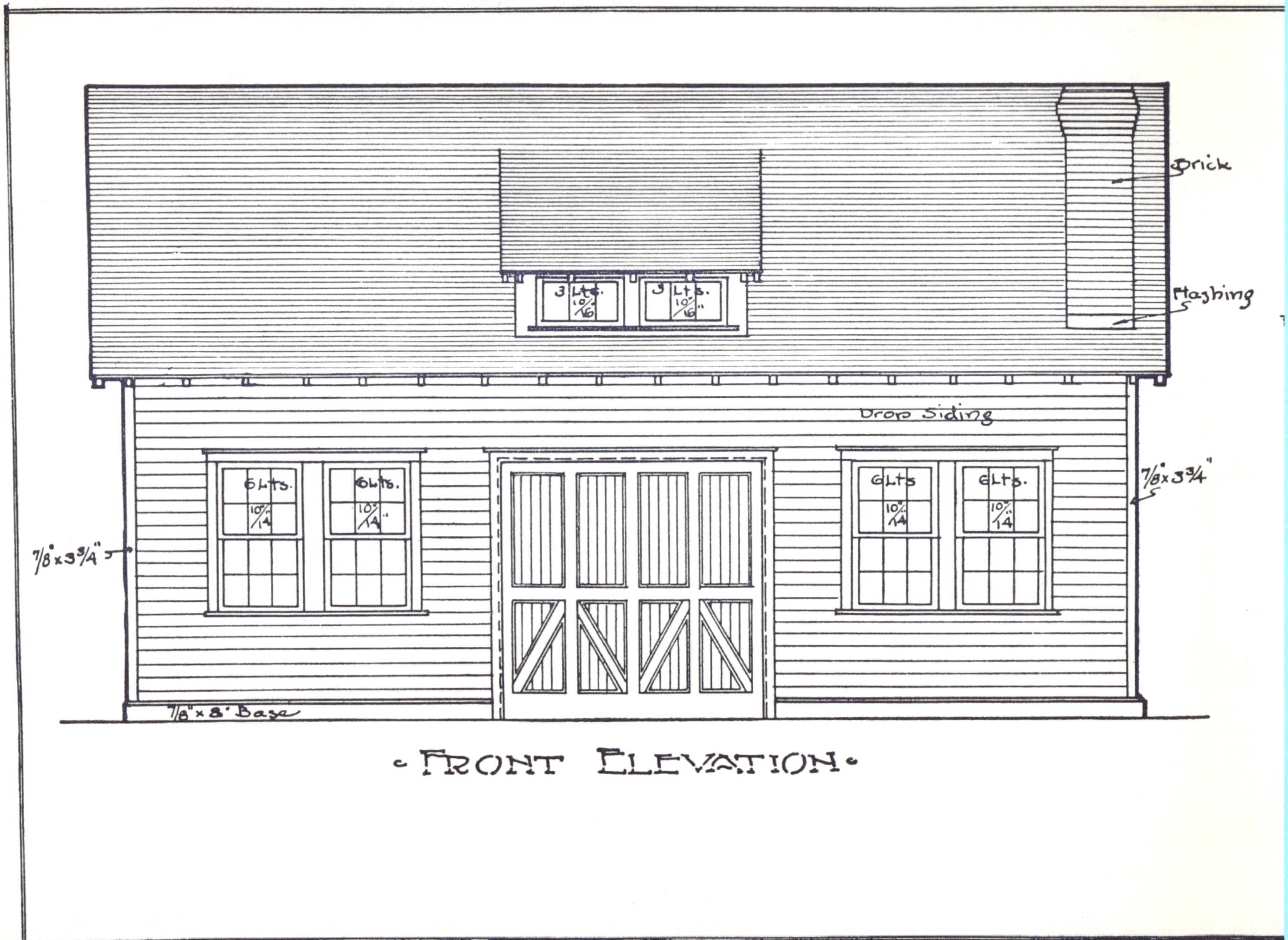

WHS149808, (both pages, split, one image)

• END ELEVATION •

• PLANS FOR •
~ MACHINE SHED ~
• INTERNATIONAL • HARVESTER •
COMPANY OF AMERICA
(INCORPORATED)

SHEET NO 1

— Scale —
0 3 6 9 12" 1-0" 2-0" 3-0" 4-0" 5-0"

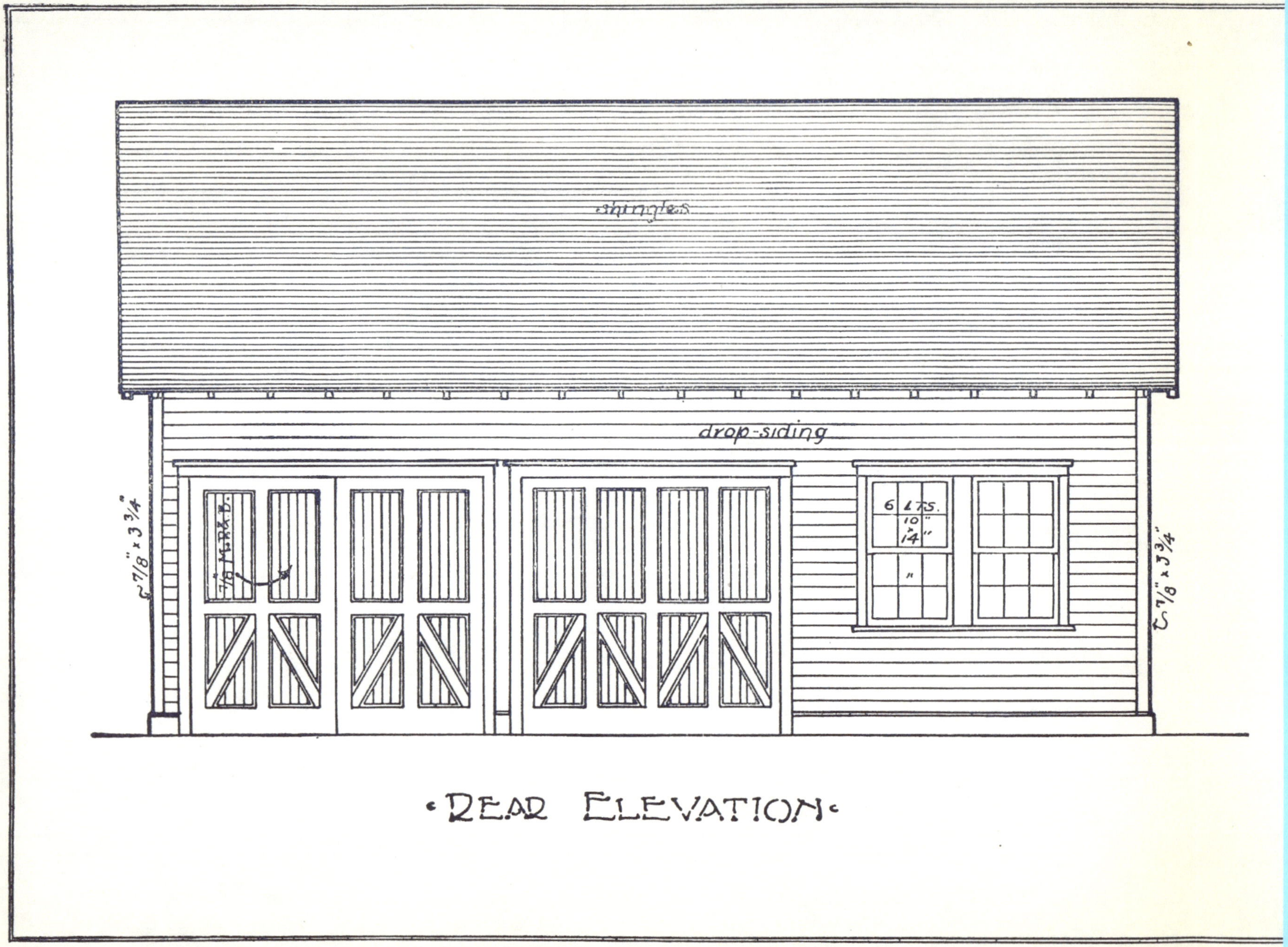

•REAR ELEVATION•

WHS150072 (both pages, split, one image)

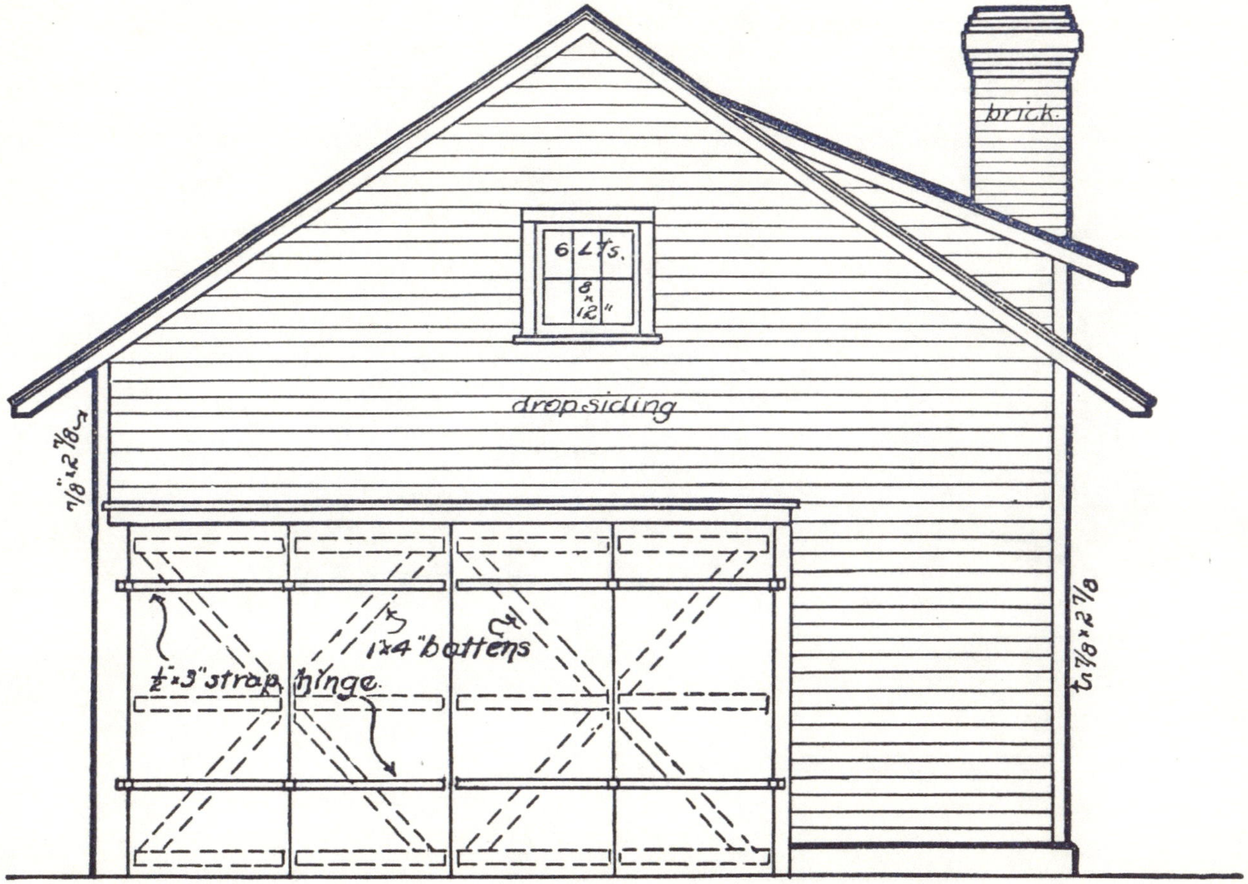

· END ELEVATION ·

· PLANS · FOR ·
· MACHINE · SHED ·
· INTERNATIONAL · HARVESTER ·
· COMPANY · OF · AMERICA ·
(INCORPORATED).

SHEET NO. 2

SCALE.
3" 6" 9" 1'-0" 2'-0" 3'-0" 4'-0" 5'-0"

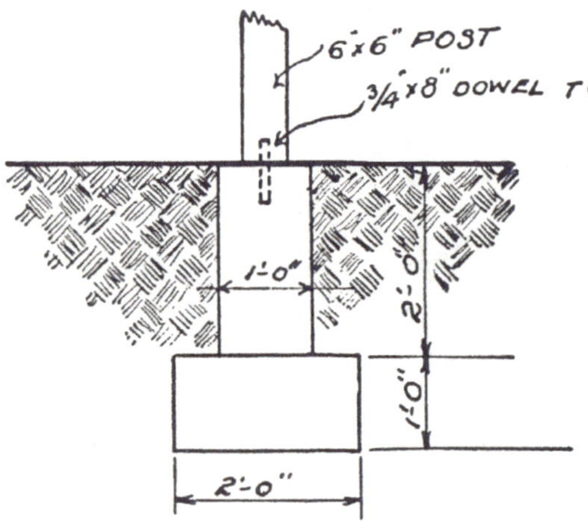

• DETAIL • OF • PIER •

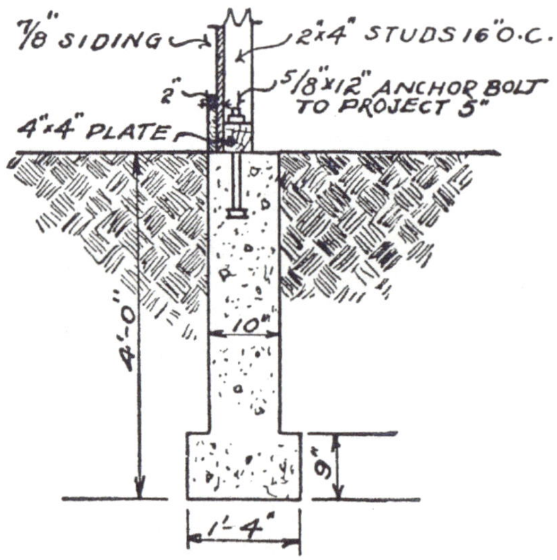

SECTION • THRU • WALL •

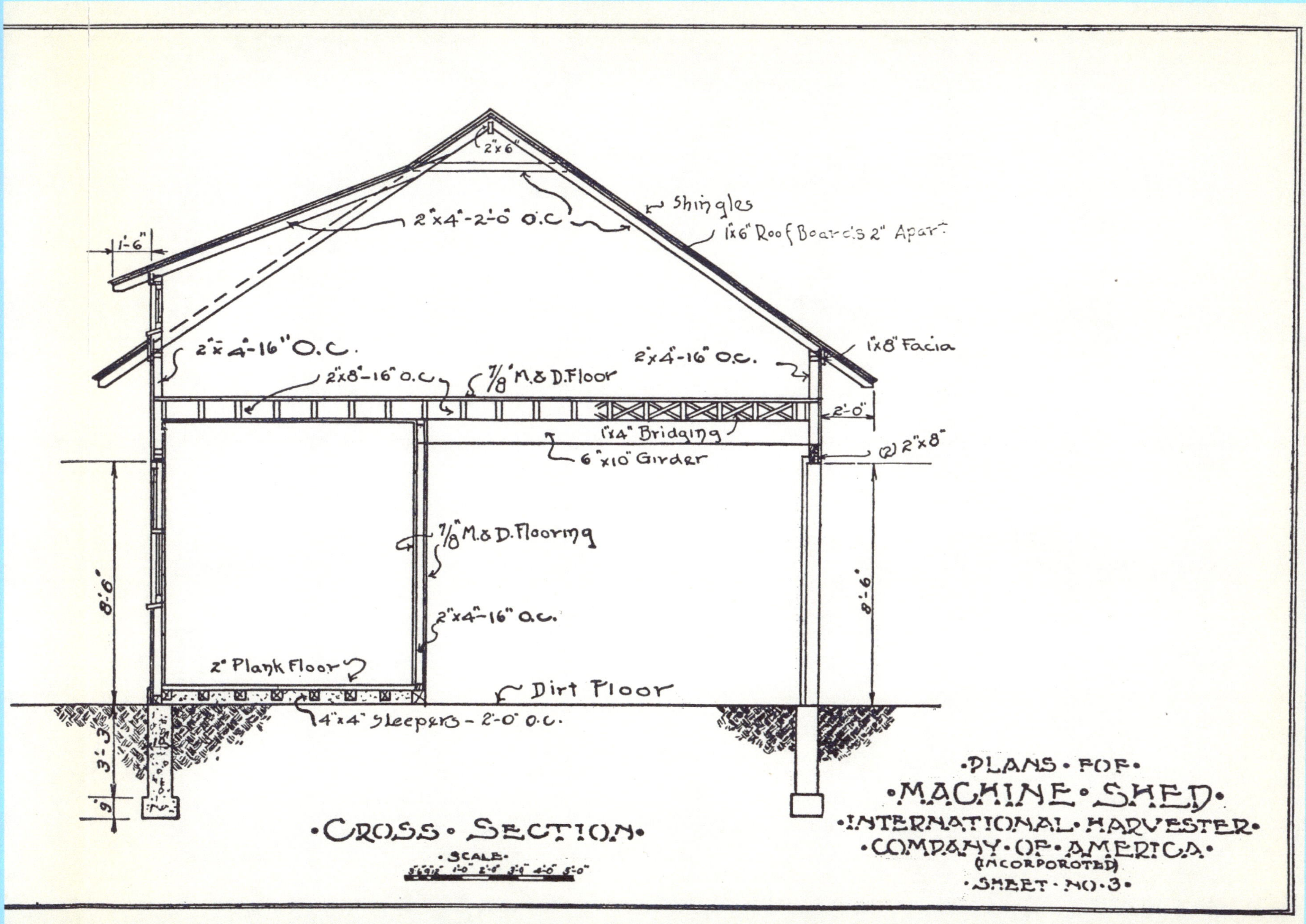
2"x6"
Shingles
2"x4"-2'-0" O.C.
1'x6" Roof Boards 2" Apart
1'-6"
2"x4"-16" O.C.
2"x4"-16" O.C.
1"x8" Facia
2"x8"-16" O.C.
7/8" M.& D. Floor
2'-0"
1"x4" Bridging
(2) 2"x8"
6"x10" Girder
7/8" M.& D. Flooring
8'-6"
8'-6"
2"x4"-16" O.C.
2" Plank Floor
Dirt Floor
4"x4" Sleepers - 2'-0" O.C.
3'-3"
9"
·CROSS·SECTION·
·SCALE·
·PLANS·FOR·
·MACHINE·SHED·
·INTERNATIONAL·HARVESTER·
·COMPANY·OF·AMERICA·
(INCORPORATED)
·SHEET·NO·3·

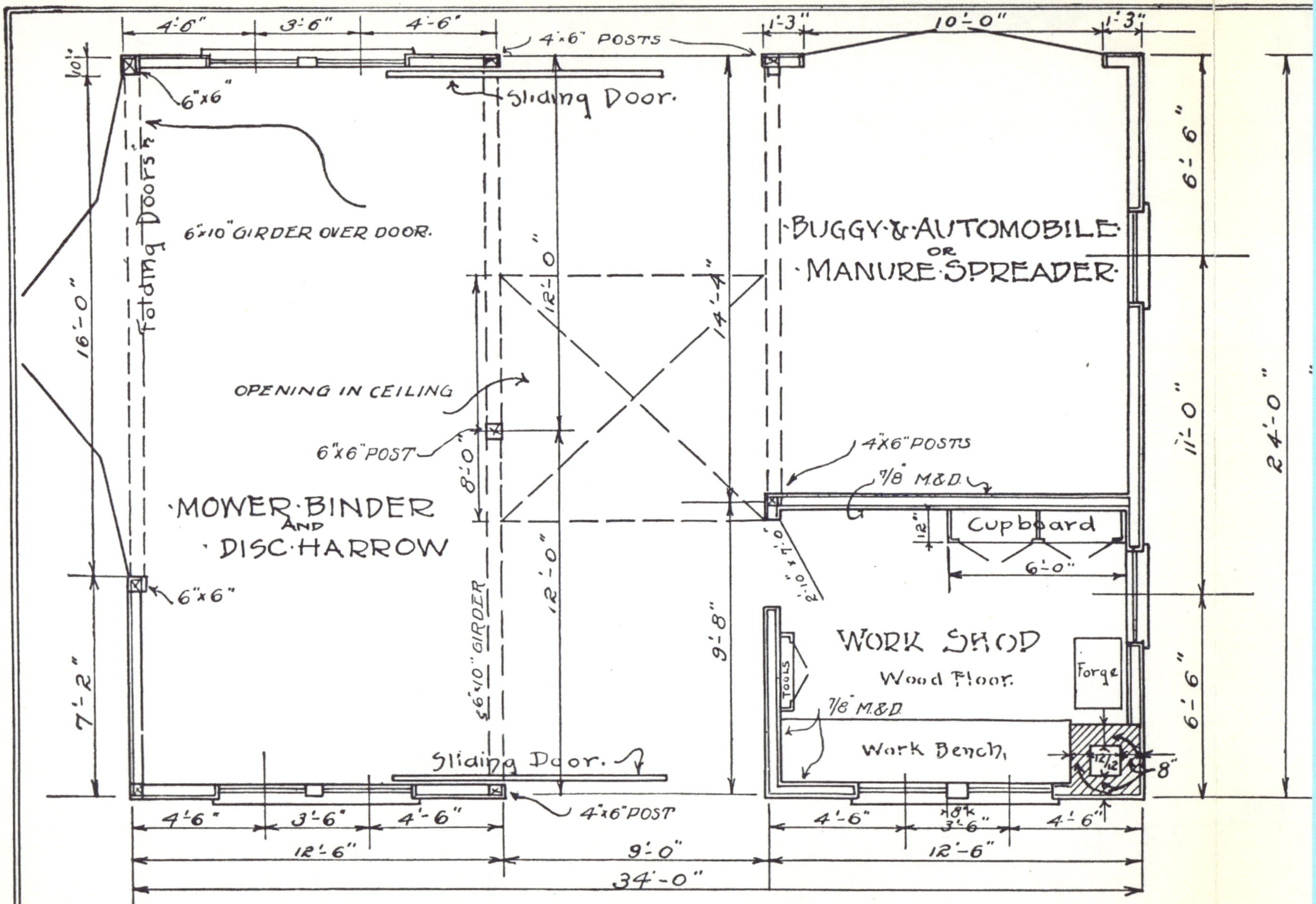

PLAN

WHS150074 (both pages, split, one image)

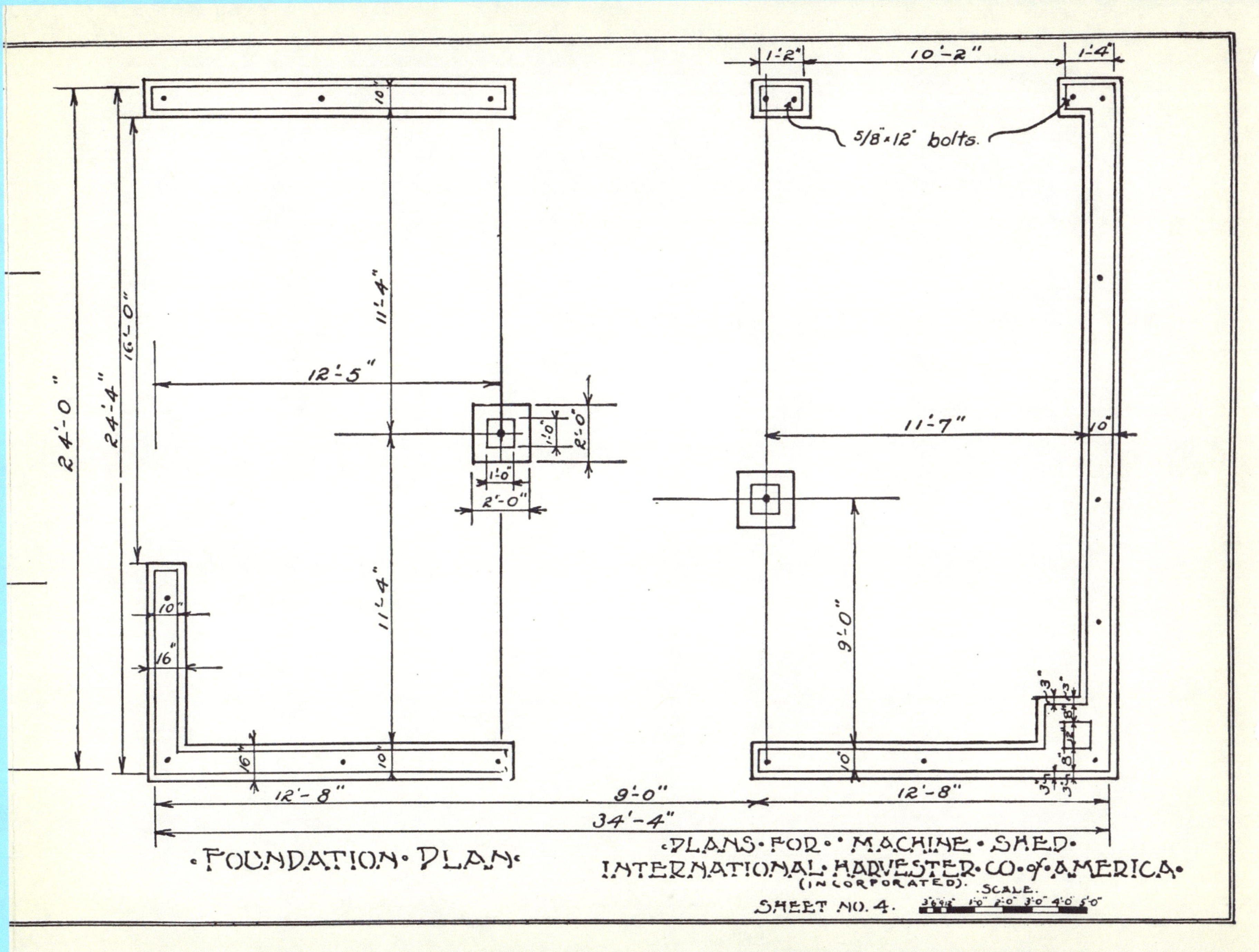

1'-2"
10'-2"
1'-4"
5/8" x 12" bolts.
24'-0"
24'-4"
16'-0"
11'-4"
12'-5"
1'-0"
2'-0"
1'-0"
2'-0"
11'-7"
10"
11'-4"
9'-0"
10"
16"
16"
10"
10"
12'-8"
9'-0"
12'-8"
34'-4"
·FOUNDATION· PLAN·
·PLANS· FOR· · MACHINE · SHED·
INTERNATIONAL· HARVESTER· CO· of· AMERICA·
(INCORPORATED).
SCALE.
SHEET NO. 4.

The following two plans are interesting with the second having a granary attached, with slightly different layouts for the different power needs.

WHS 147493

BUILDING PLANS SPECIFICATIONS AND MATERIAL LIST FOR A FARM POWER HOUSE

ISSUED AND COPYRIGHTED 1912 BY

INTERNATIONAL HARVESTER COMPANY OF AMERICA
(INCORPORATED)
CHICAGO USA

International Harvester gave suggestions in the beginning of every building plan, points to remember and layout suggestions. Some of which are as follows:

1) The engine in a power house is the most important machine and should be installed in a room that can be closed off from the remainder of the rooms, and placed so that all sides are accessible, at least three feet from any wall. The gasoline tank should be located outside, about 30 feet distant, in the ground, to insure against fire risk and kept cool in summer so that the gasoline will not evaporate.

2) In the engine room might be located the grindstone, emery wheel, drill and work bench, this room then becomes a workshop. If a dynamo is used to furnish light for the house, it can also be located in this room.

3) The remaining space should be divided into two rooms. One can locate the grinder, sheller, cutter and fanning mill. The grinder and sheller should be close to the window so it is possible to unload the wagon through the window into the machine. This room should be tightly partitioned off from the others because these machines create dust, which would interfere with the successful operation of the other machines.

4) In the third room can be located the churn, cream separator, pump and drain, dairy table, and large washing trough. This room becomes the dairy and should be shut off from the remainder of the power house by dust tight partitions.

5) All these machines in the power house should be driven by belt from a line shaft set overhead or hung from the ceiling. All high speed machines should be firmly bolted to the floor and operated by their own belts.

6) The line shafting can be extended through the end of the building for that a blacksmith bellows or circular saw or any other wood working machinery can be driven on the outside from the same power. When an engine of 4-horsepower or less is installed, 1 1/2” iron shafting will be adequate. If a larger engine is installed, 2 inch iron shafting should be used.

7) The following plans have been carefully worked out and show the ideal locations of the different machines. The cost of this farm powerhouse, without machines, based on the price of material and labor in Illinois would be approximately $450. This figure can be reduced by decreasing the dimensions.

SHEET 1. Scale, ¼ inch to 1 foot

Farm Power House Building Plans

International Harvester Company of America (Incorporated)

Elevation - CC

WHS150075

SPECIFICATIONS AND MATERIAL LIST

TAKE THIS LIST TO YOUR DEALER IN BUILDING MATERIAL AND ASK HIM TO GIVE YOU THE COST.
ANY CARPENTER CAN GIVE YOU THE COST OF THE LABOR

FOUNDATION

STONE

1¾ Cords Stone
10 Bushels Lime
2 Yards Sand

CONCRETE

6 Bbls. Portland Cement
6 Yards good coarse Gravel

CHIMNEY

400 Hard Burned Brick
½ Yard Sand
3 Bags Cement

4" CONCRETE FLOOR

6 Bbls. Portland Cement
6 Yards good coarse Gravel

WOOD FLOOR

2 Pieces 4" x 6"—16'-0" No. 1 Common Lumber Side Sills
2 " 4" x 6"—14'-0" No. 1 " " End Sills
2 " 4" x 6"—14'-0" No. 1 " " Center Support
2 " 4" x 6"—14'-0" No. 1 " " Side Sills
21 " 2" x 6"—14'-0" No. 1 " " Joists
450 feet 1" x 8 or 10"-12'-0" to 16'-0" No. 1 Com. S. L. S. under Floor
500 feet ⅞" x 2¼" Face No. 1 Factory Maple Flooring

WALLS

4 Pieces 2" x 4"—14'-0" No. 1 Common Lumber Side Sills
2 " 2" x 4"—14'-0" No. 1 " " End Sills
2 " 2" x 2"—14'-0" No. 1 " " Partition Sills
34 " 2" x 4"—10'-0" No. 1 " " Wall Studs
8 " 4" x 4"—10'-0" No. 1 " " " "
2 " 1" x 4"—10'-6" No. 1 " " " "
2 " 2" x 4"—16'-0" No. 1 " " " "
8 " 2" x 4"—18'-0" No. 1 " " " "
4 " 2" x 4"—14'-0" No. 1 " " " "
1 " 2" x 4"—14'-0" No. 1 " " " "
2 " 2" x 4"—16'-0" No. 1 " " " "

ROOF

19 Pieces 2" x 4"—18'-0" No. 1 Common Lumber Rafters
7 " 2" x 4"—24'-0" No. 1 " " Collar Beams
7 " 2" x 4"—14'-0" No. 1 " " " "
4 " 2" x 4"—16'-0" No. 1 " " Plates
4 " 2" x 4"—20'-0" No. 1 " " "
4 " 2" x 4"—14'-0" No. 1 " " "
800 Ft. Roof Boards
800 Sq. Ft. Prepared Roofing, or, 26 Bunches of Shingles

MISCELLANEOUS

2 Posts 4" x 4"— 9'-10" No. 1 Common Lumber
2 Pieces 3" x 8"—28'-0" Bridging
12 " 1" x 4"—20'-0" No. 1 Common Lumber
14 " 1" x 4"—18'-0" No. 1 " "
2 " 1" x 4"—14'-0" No. 1 " "
3 " 1" x 6"—20'-0" No. 1 " "
5 " 1" x 6"—18'-0" No. 1 " "
8 " 1" x 6"—16'-0" No. 1 " "
2 " 1" x 8"—18'-0" No. 1 " "
2 " 2" x 8"—16'-0" Window Sill
124 Lineal Feet Window Stops
124 " " Parting "
1 Piece 2" x 8"—8'-0" Oak Door Sills
54 Lineal Feet Door Stops
2 Pieces 2" x 12"—10'-0" Work Bench
1 Piece 2" x 12"—2'-0" Support Under Chimney
3 Pieces 1" x 12"—8'-0" " " " and Shelves
4 Doors, 3'-0" x 7'—1¾"
8 Windows, 10" x 16", 12 Lights
1200 Feet No. 1 Matched Fencing or Drop Siding
800 " No. 1 " " for Partitions and Ceiling

HARDWARE

8 Loose Pin Door Butts, 5" x 5"
4 Door Locks
32 Sash Bolts
2 Iron Pins, 1" Dia. 9" Long
50 Lbs. 20d Wire Nails
50 Lbs. 10d " "
100 Lbs. 8d " "
25 Lbs. 3d Fine Wire Nails—if Shingles are used
30'-0—1$\frac{15}{16}$" Dia. Cold Rolled Shafting
1 Compression Coupling
7 18" Drop Hangers

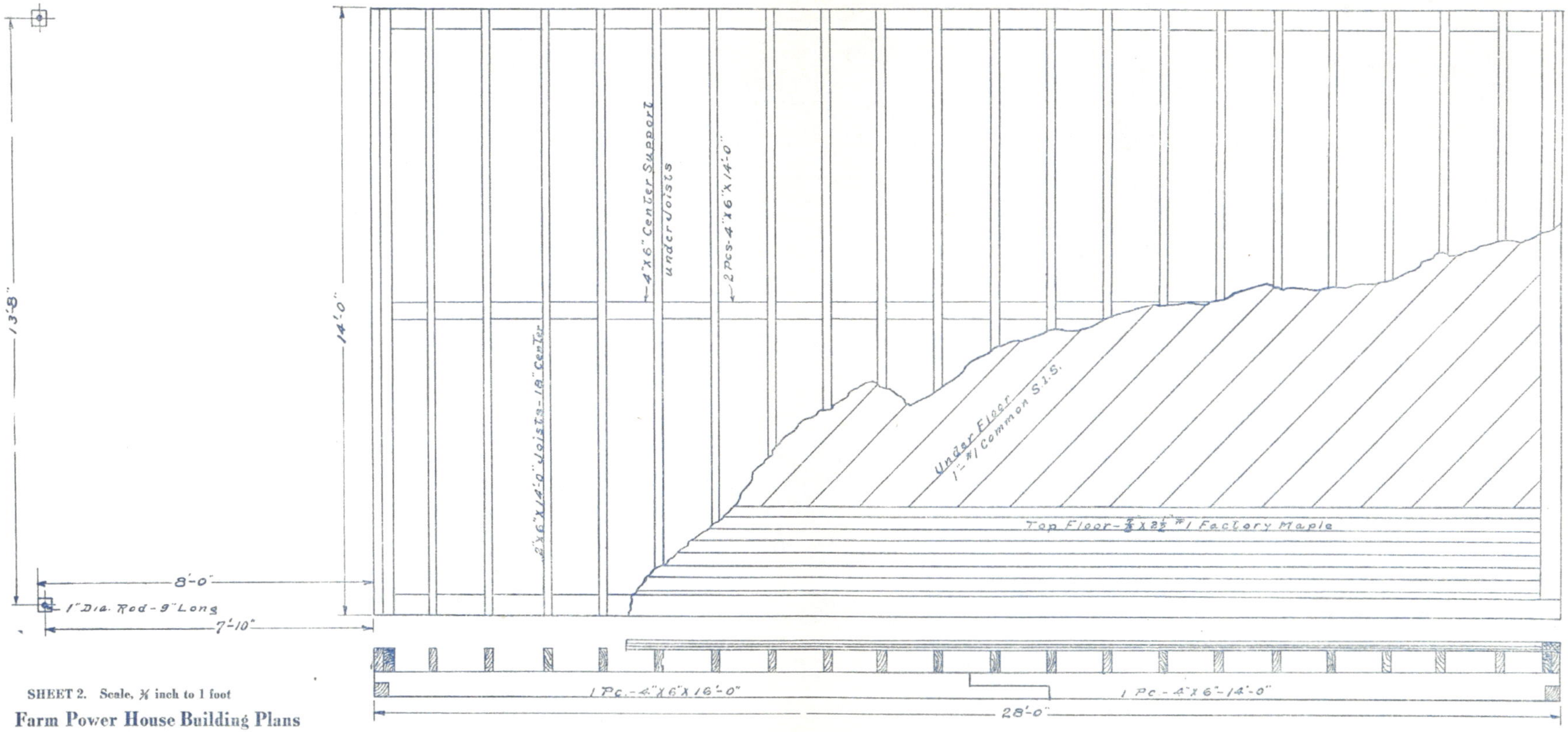

SHEET 2. Scale, 3/8 inch to 1 foot

Farm Power House Building Plans

International Harvester Company of America
(Incorporated)

Plan of Sills, Joists and Floors

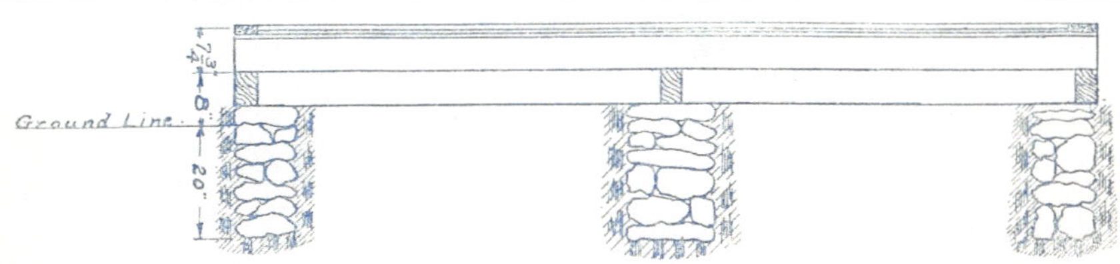

WHS150076, same image, split

SHEET 3. Scale, ¼ inch to 1 foot

Farm Power House Building Plans

International Harvester Company of America
(Incorporated)

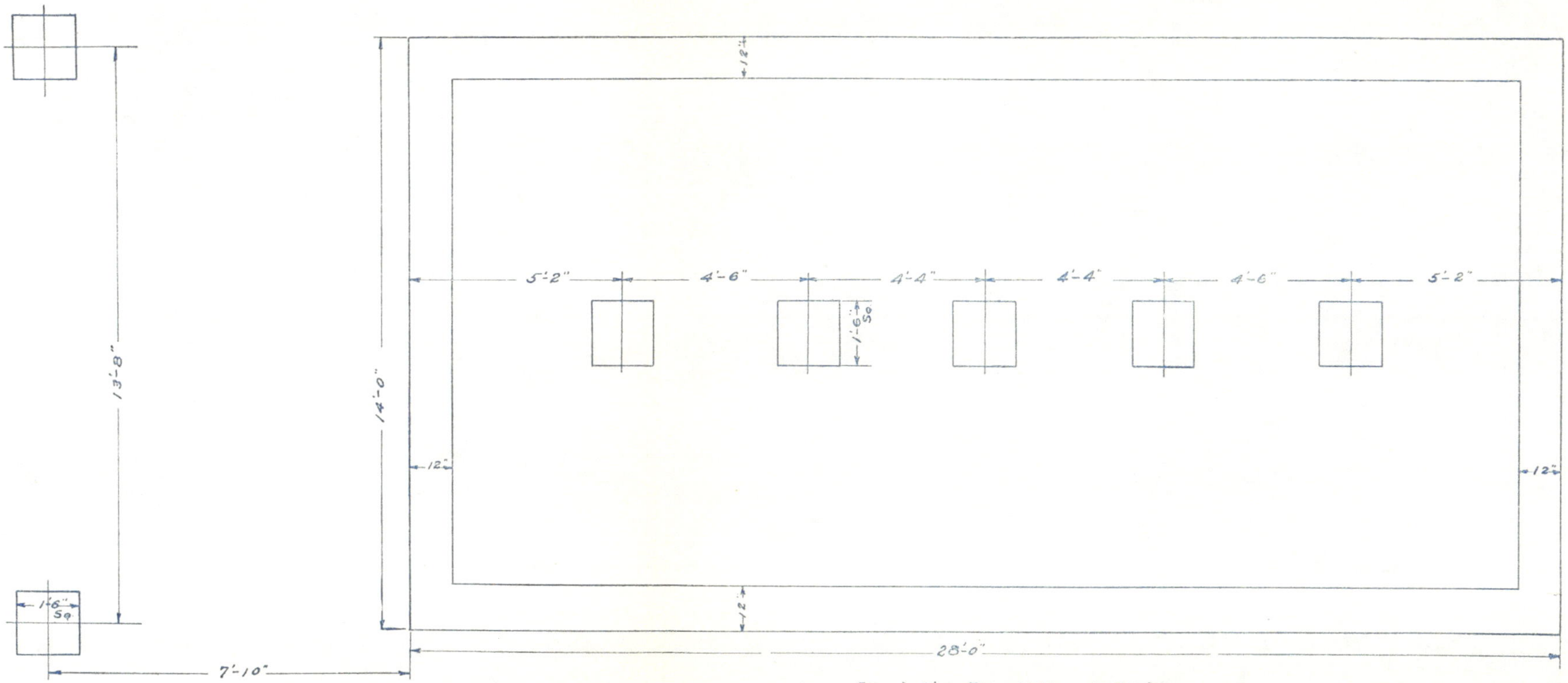

Foundation Plan if Stone Is Used

WHS150077

An I H C Engine installed to provide a practical Farm Power House

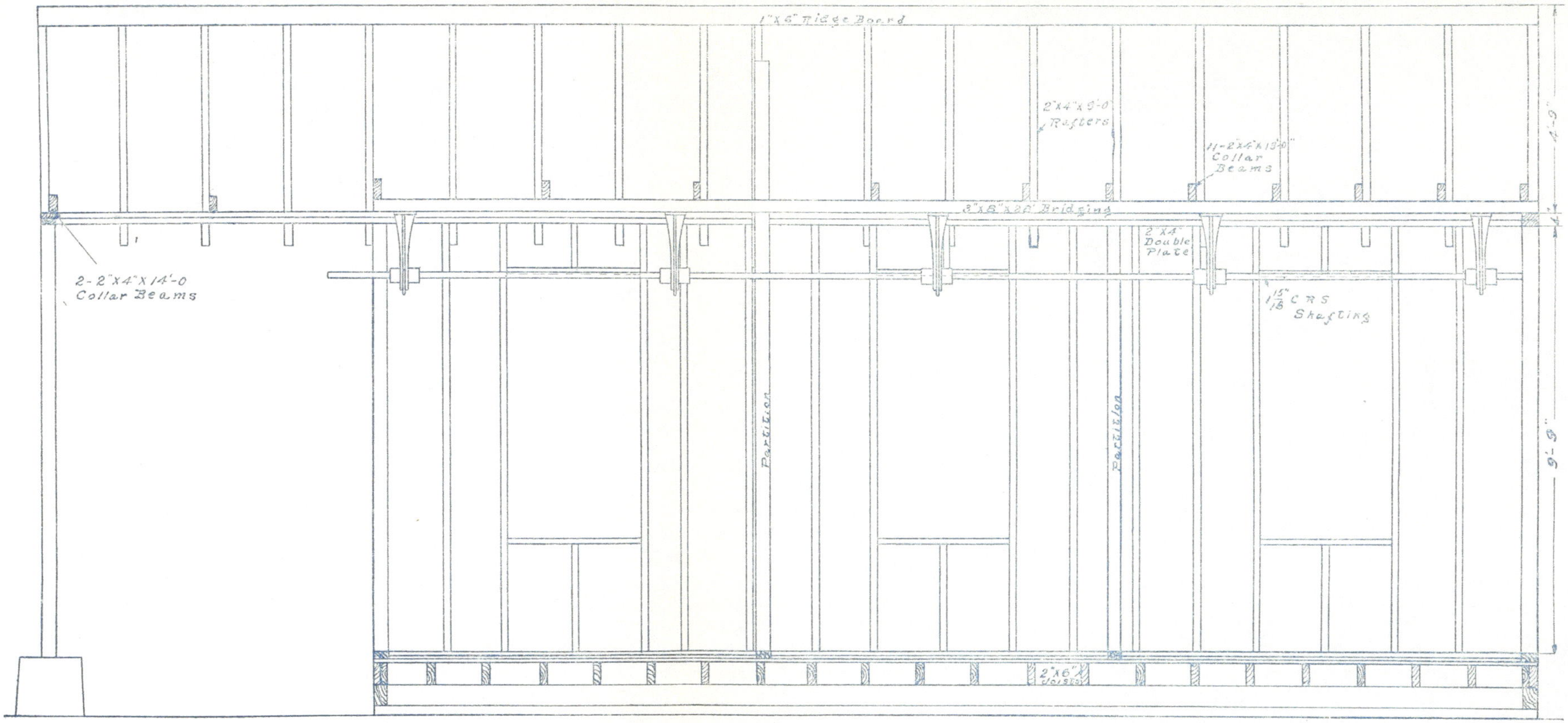

Longitudinal Section

SHEET 4 Scale, ¼ inch to 1 foot

Farm Power House Building Plans

International Harvester Company of America
(Incorporated)

WHS150078

WHS150079, one image, split pages

End Elevation. - BB

SHEET 5. Scale, ⅜ inch to 1 foot

Farm Power House Building Plans

International Harvester Company of America
(Incorporated)

End Elevation - A A

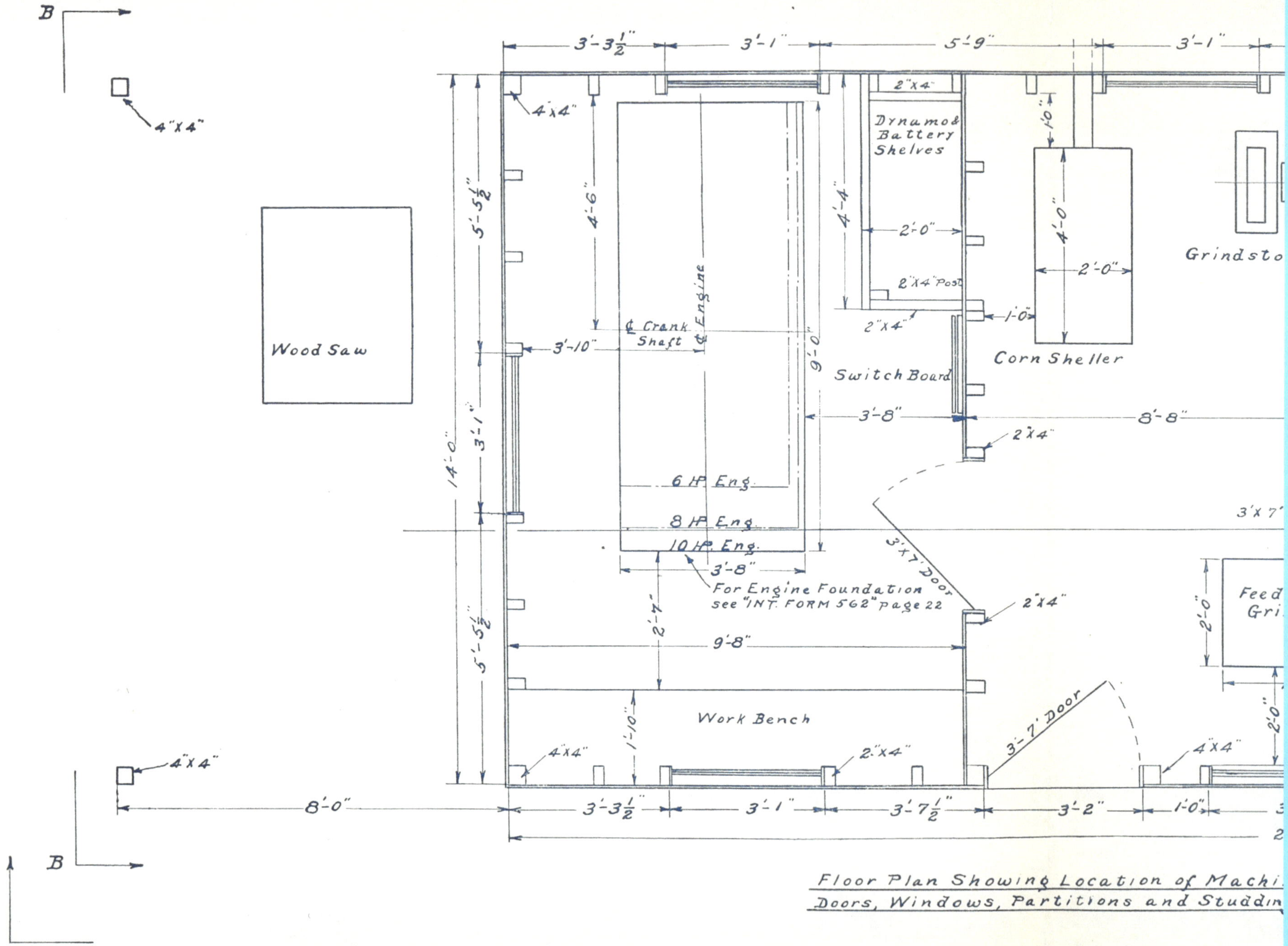

Floor Plan Showing Location of Machi
Doors, Windows, Partitions and Studdin

WHS150081, (One image, split page)

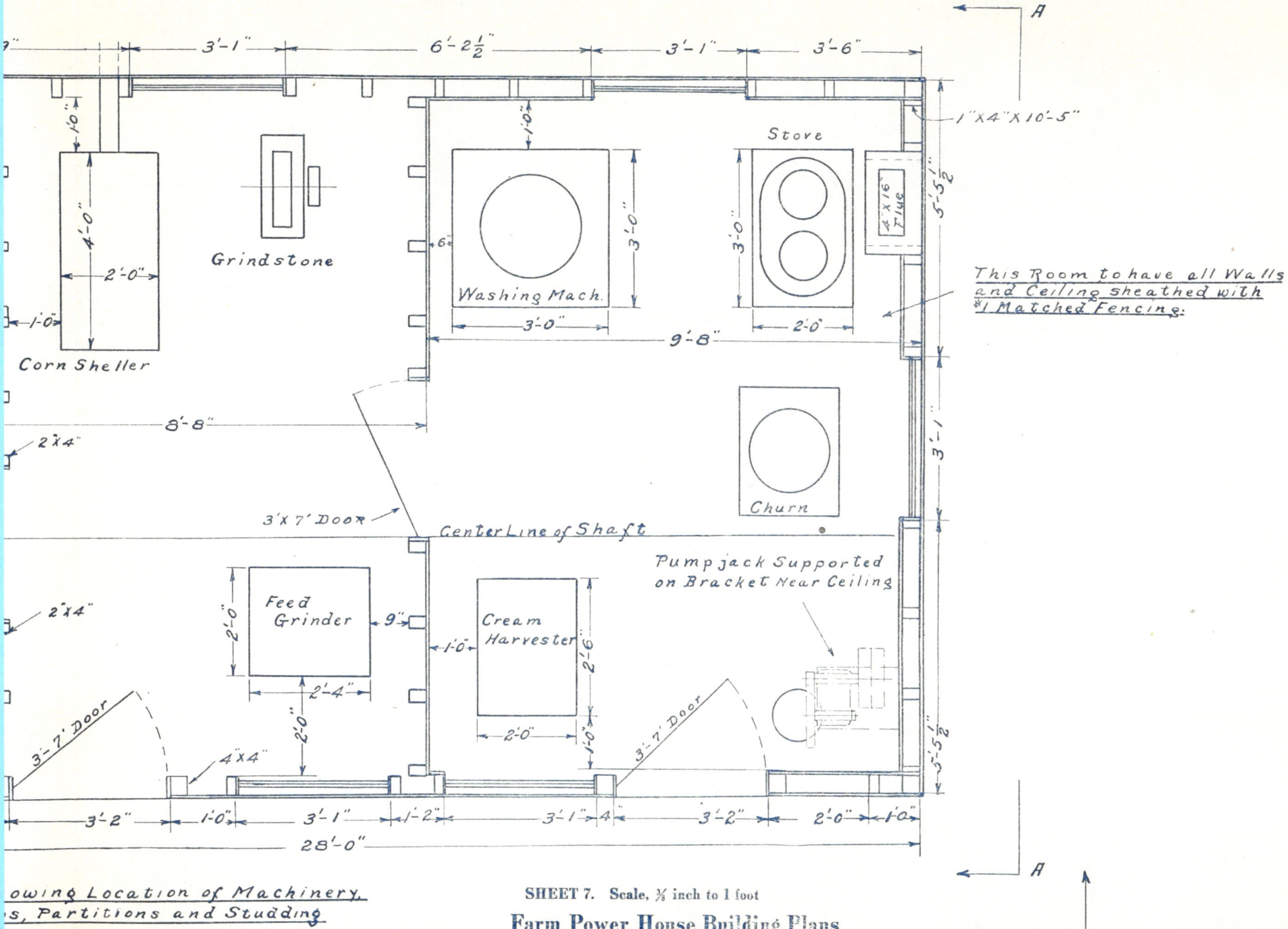

SHEET 7. Scale, ⅜ inch to 1 foot

Farm Power House Building Plans

International Harvester Company of America
(Incorporated)

Farms run more efficiently when power is available, and in the early 1900s power was obtained by stationary engines, kept in a separate building, called the powerhouse. This plan has a granary attached, for ease of processing grains for animals or for sale.
WHS147494

BUILDING PLANS SPECIFICATIONS AND MATERIAL LIST FOR A FARM POWER HOUSE IN CONNECTION WITH GRANARY

ISSUED AND COPYRIGHTED 1912 BY

INTERNATIONAL HARVESTER COMPANY OF AMERICA
(INCORPORATED)
CHICAGO USA

566

SHEET 1. Scale, ⅜ inch to 1 foot
Building Plans
Farm Power House with Granary
International Harvester Company of America
(Incorporated)
36'-0⅞"
ELEVATION -C.C.

WHS150126

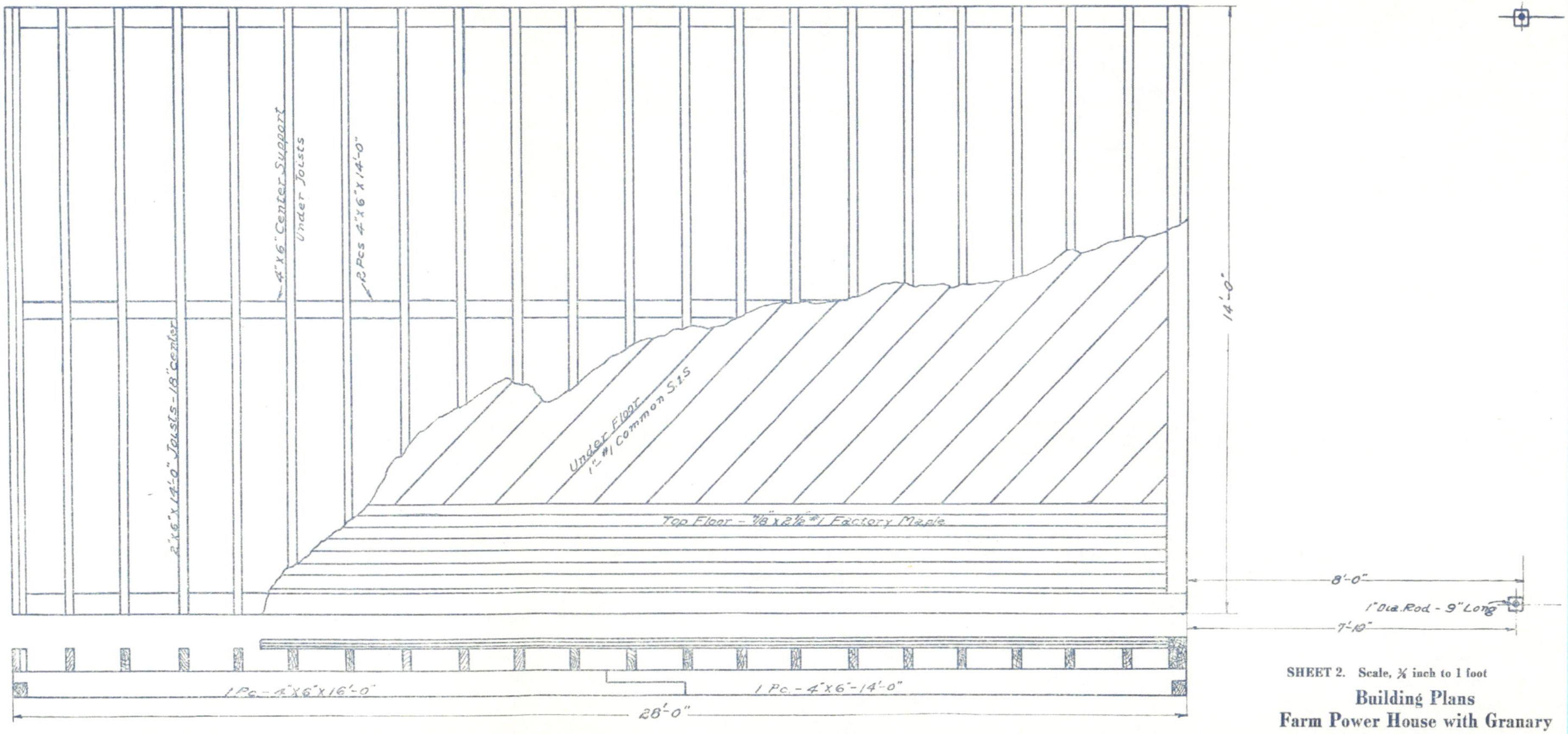

Plan of Sills, Joists and Floors

SHEET 2. Scale, ⅜ inch to 1 foot

Building Plans

Farm Power House with Granary

International Harvester Company of America (Incorporated)

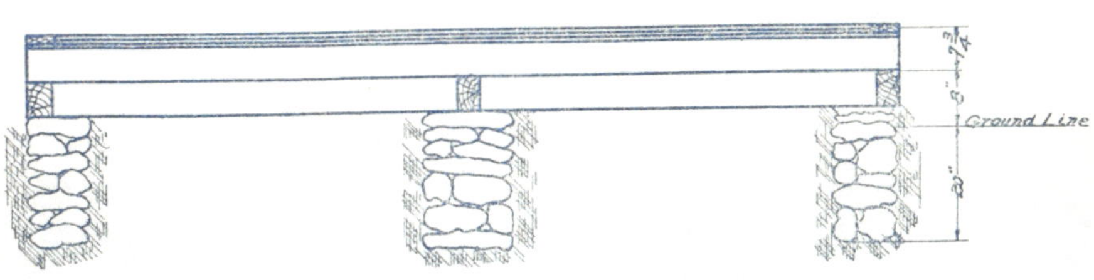

WHS150140, one image, split

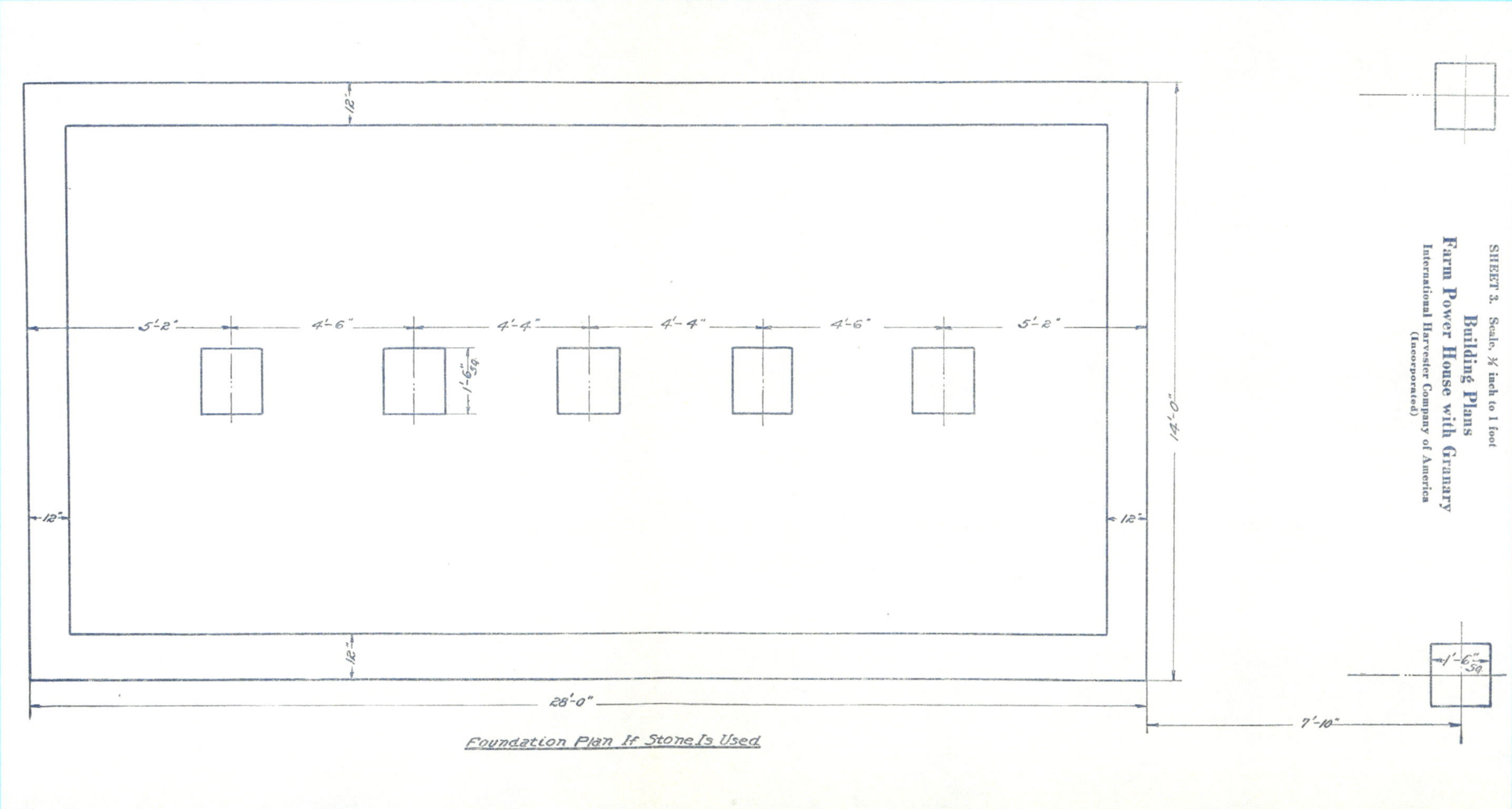
SHEET 3. Scale, ¼ inch to 1 foot
Building Plans
Farm Power House with Granary
International Harvester Company of America
(Incorporated)
12"
5'-2"
4'-6"
4'-4"
4'-4"
4'-6"
5'-2"
1'-6" sq
14'-0"
12"
12"
12"
28'-0"
1'-6" sq
7'-10"
Foundation Plan If Stone Is Used

WHS150141

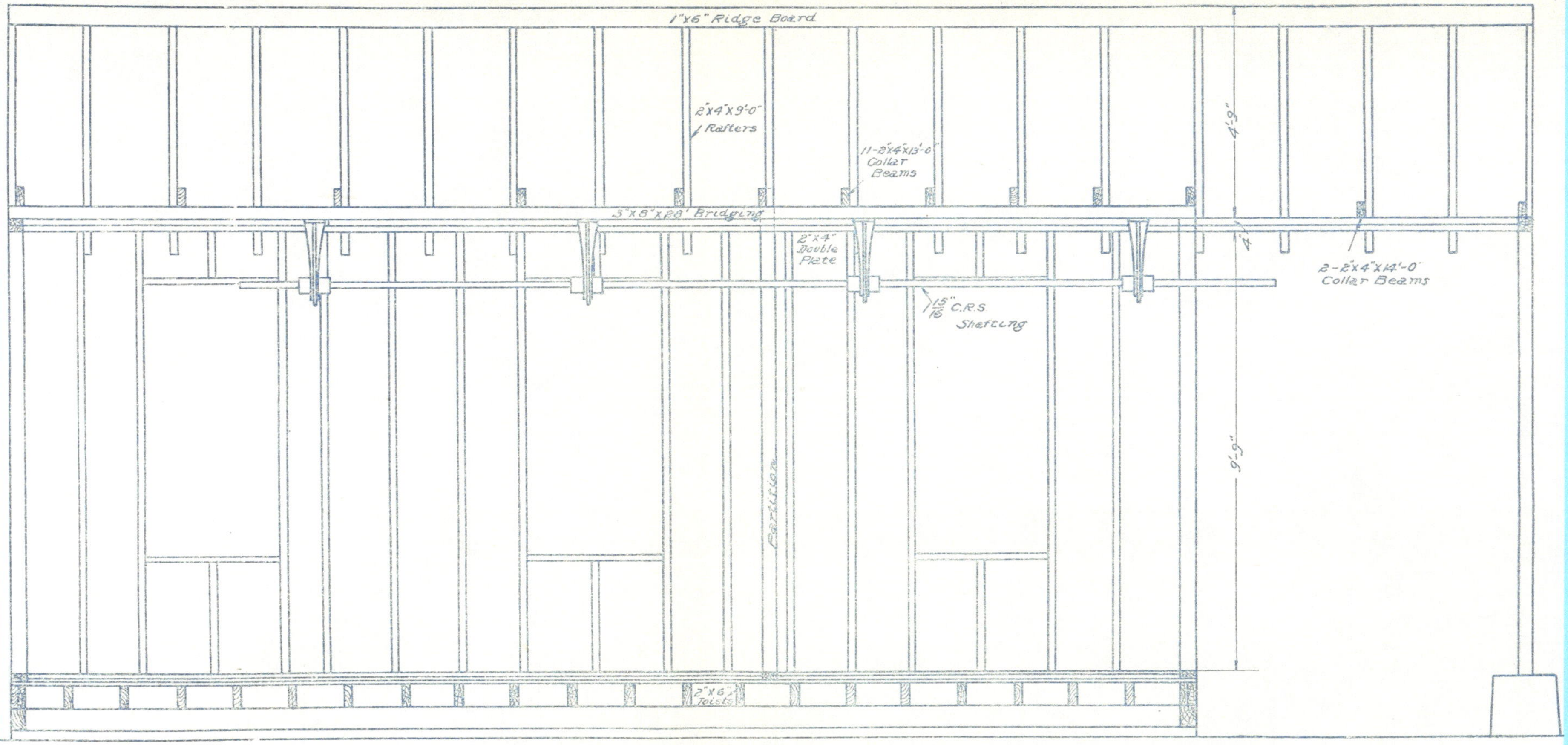

Longitudinal Section

SHEET 4. Scale, ⅜ inch to 1 foot

Building Plans
Farm Power House with Granary

International Harvester Company of America
(Incorporated)

WHS150145, one image, split page

SHEET 5. Scale, ⅜ inch to 1 foot

Building Plans
Farm Power House with Granary

International Harvester Company of America
(Incorporated)

END ELEVATION - A-A

WHS150147 one image, split page

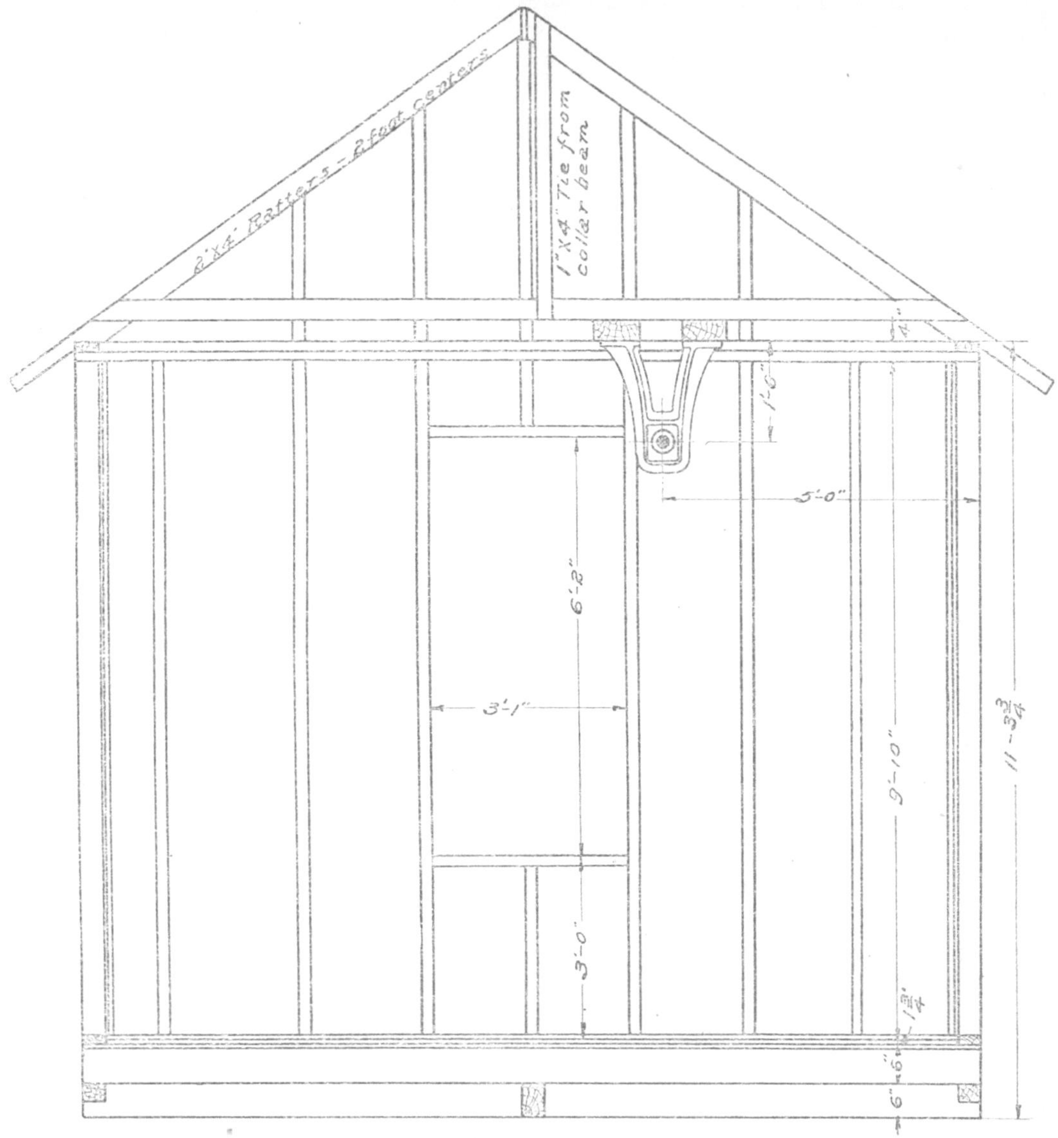

Framing of Building End

SHEET 6. Scale, ⅜ inch to 1 foot

Building Plans
Farm Power House with Granary

International Harvester Company of America
(Incorporated)

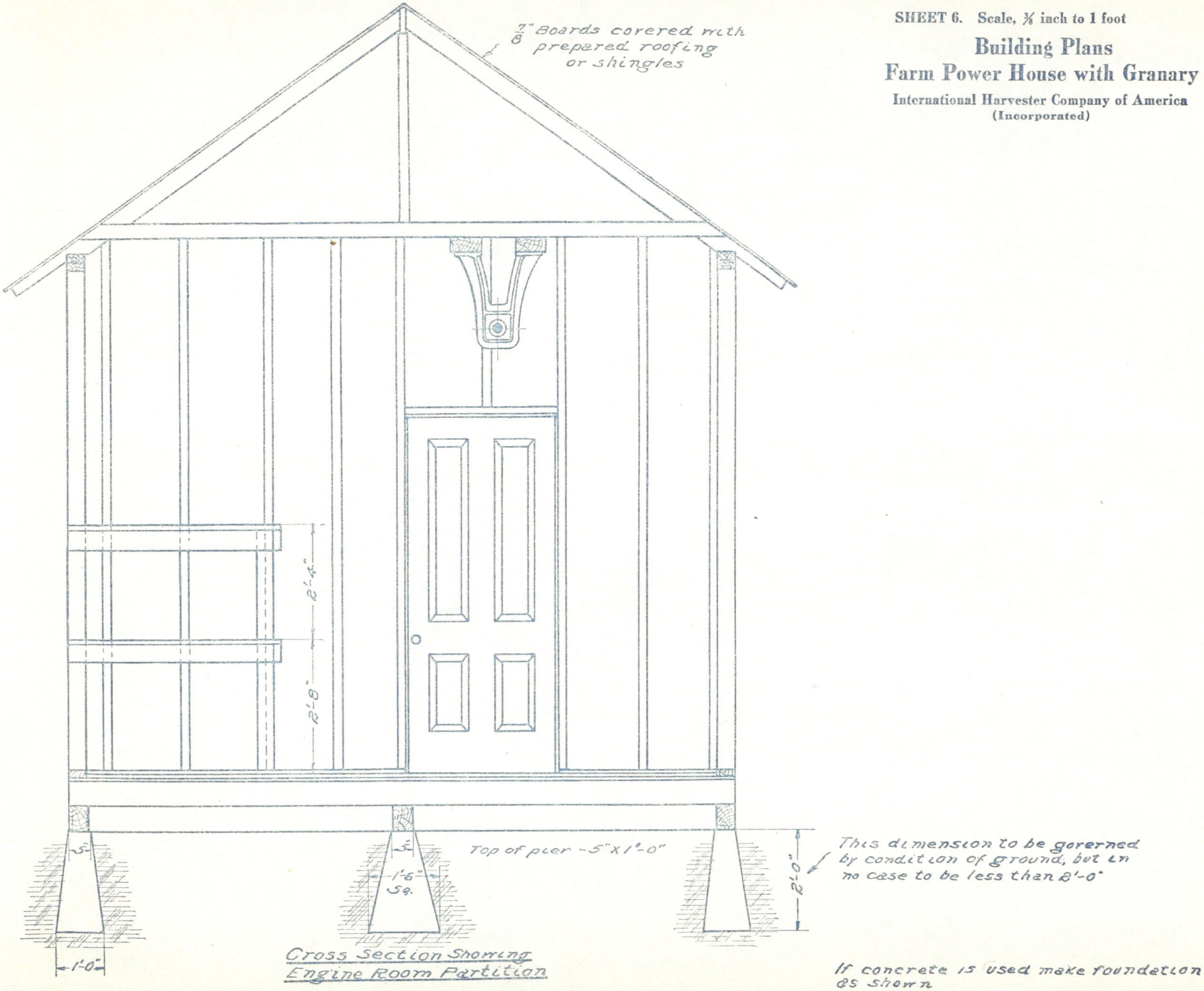

Cross Section Showing Engine Room Partition

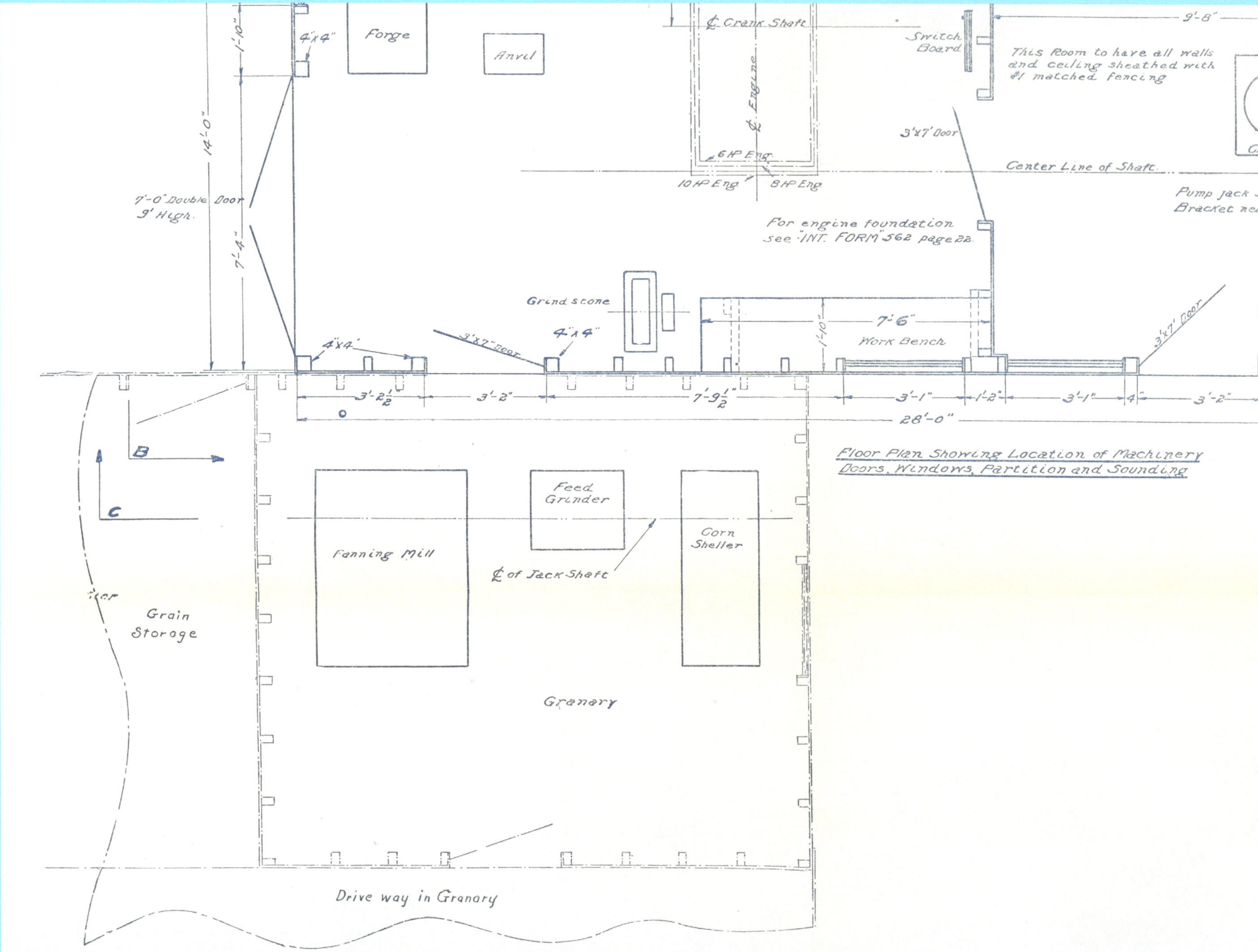

 WHS150149, One image, split page

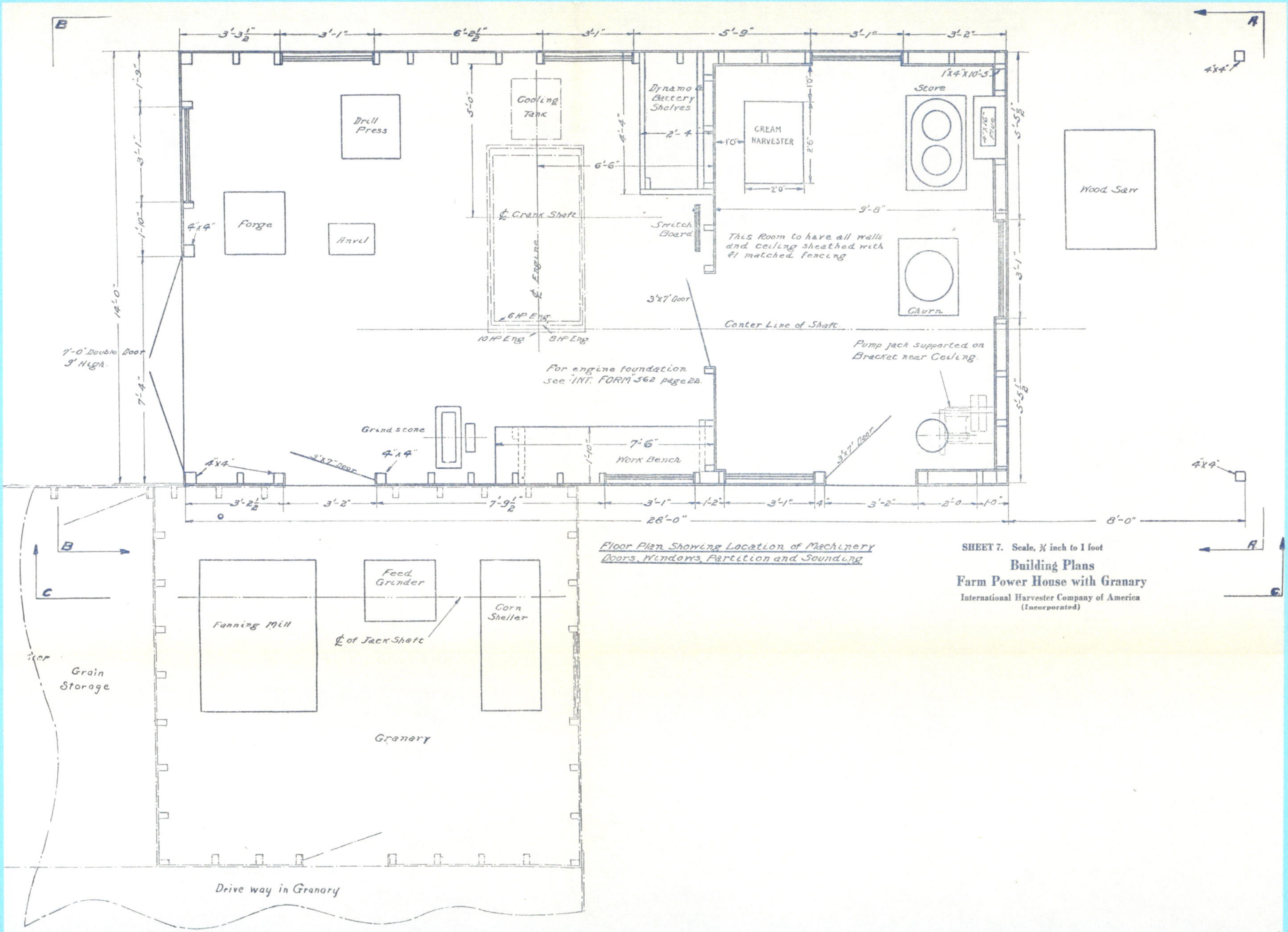

SHEET 7. Scale, ¼ inch to 1 foot

Building Plans
Farm Power House with Granary

International Harvester Company of America
(Incorporated)

BUILDING PLANS SPECIFICATIONS AND MATERIAL LIST FOR A HOG HOUSE

ISSUED AND COPYRIGHTED 1912 BY

INTERNATIONAL HARVESTER COMPANY OF AMERICA
(INCORPORATED)
CHICAGO USA

Pork is big business, has been for many years, as viewed with this layout which many piggeries still employ the same concepts today.

WHS147498

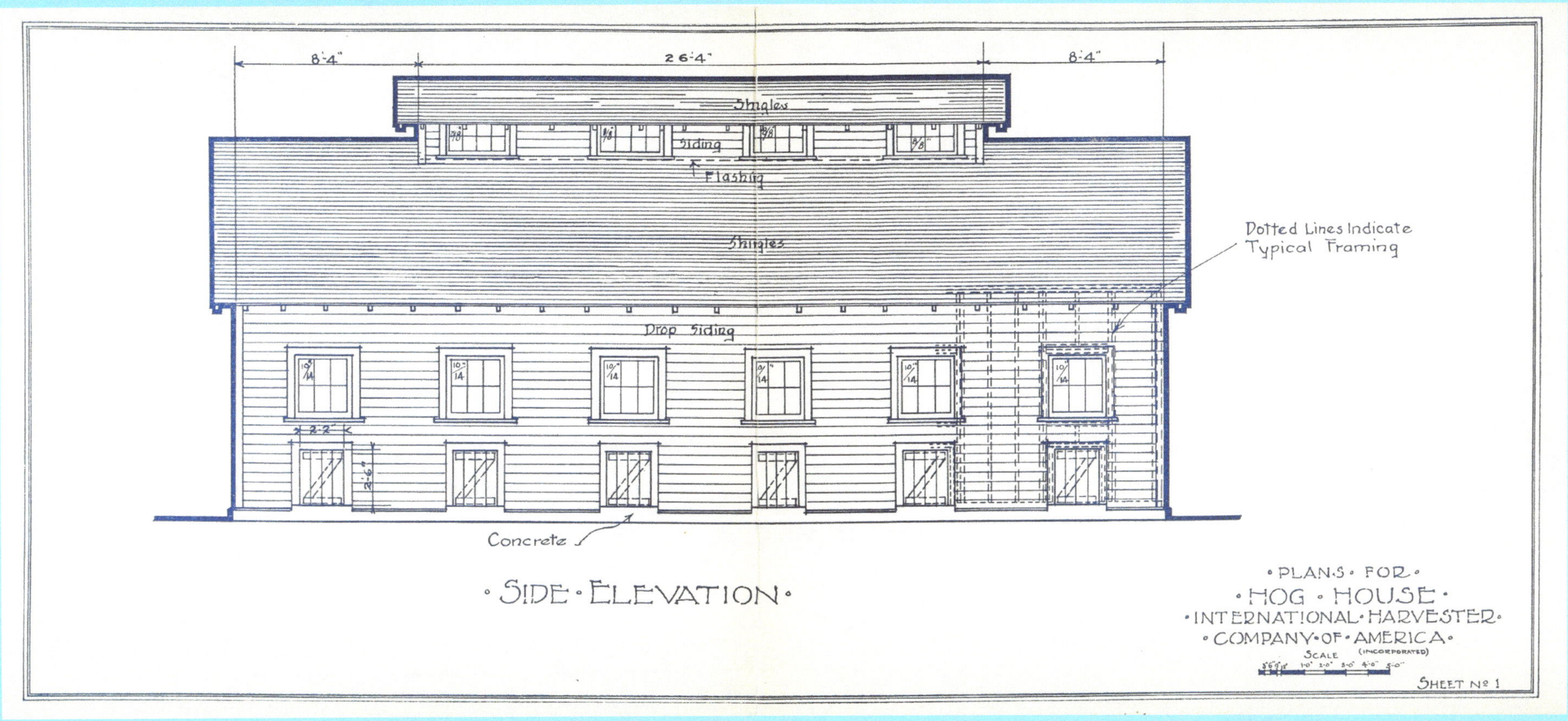
8'-4"
26'-4"
8'-4"
Shingles
Siding
Flashing
Shingles
Dotted Lines Indicate Typical Framing
Drop Siding
2'-2"
Concrete
• SIDE • ELEVATION •
• PLANS • FOR •
• HOG • HOUSE •
• INTERNATIONAL • HARVESTER •
• COMPANY • OF • AMERICA •
(INCORPORATED)
SCALE
SHEET Nº 1

WHS150121

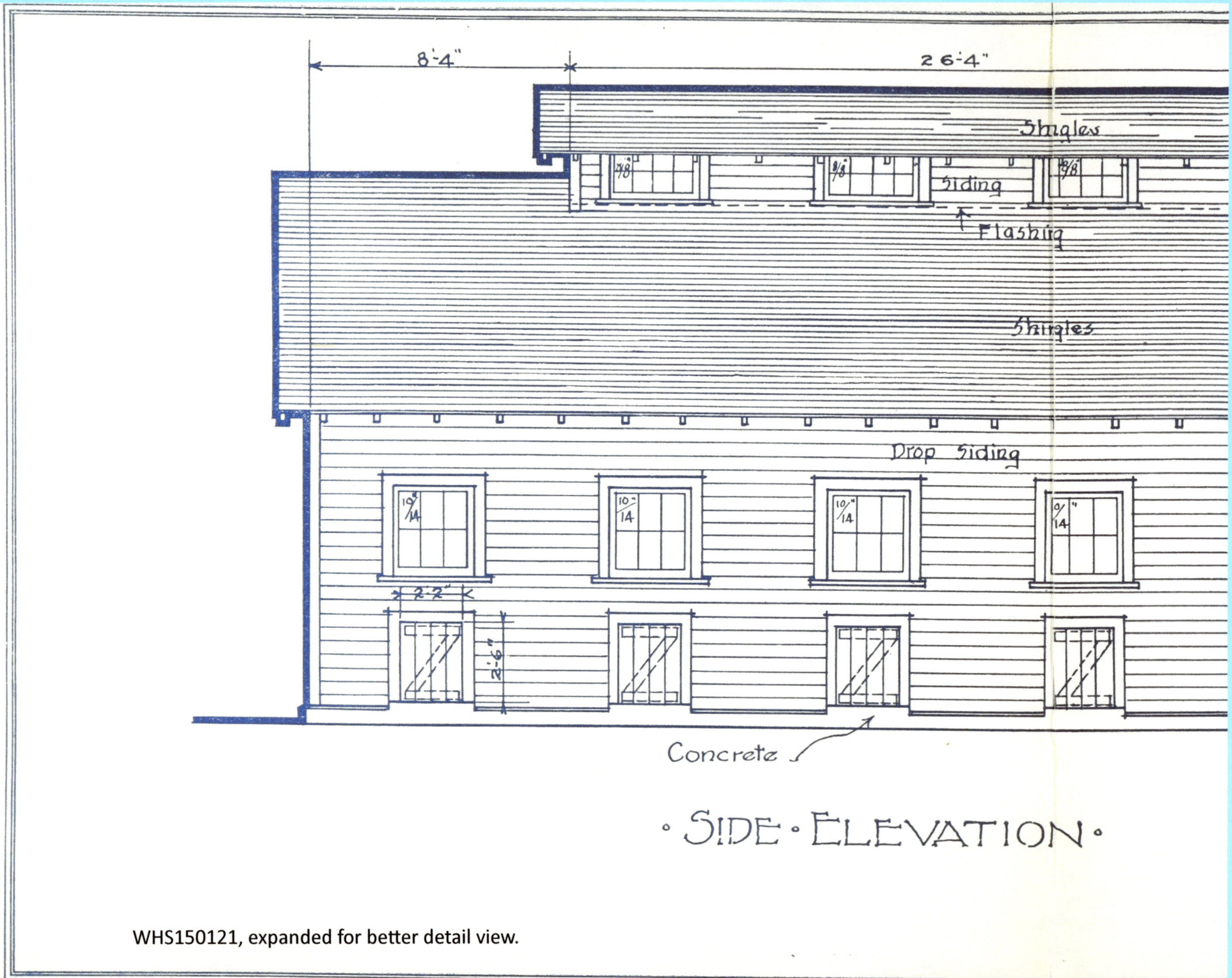

WHS150121, expanded for better detail view.

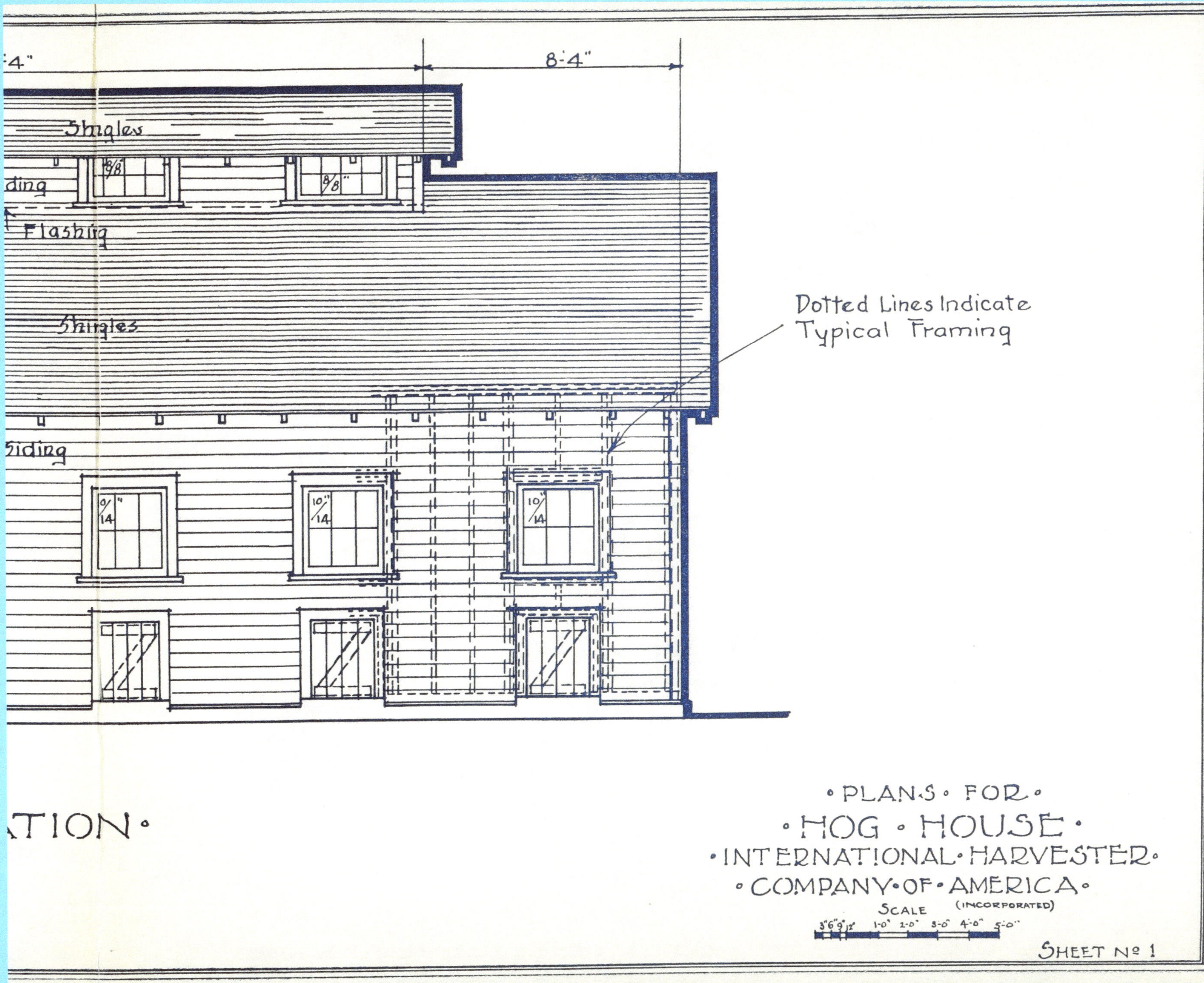

8'-4"
Shingles
8/8
8/8"
ding
Flashing
Shingles
Dotted Lines Indicate
Typical Framing
Siding
10"/14
10"/14
10"/14
ATION
• PLANS • FOR •
• HOG • HOUSE •
• INTERNATIONAL • HARVESTER •
• COMPANY • OF • AMERICA •
(INCORPORATED)
SCALE
3" 6" 9" 12" 1'-0" 2'-0" 3'-0" 4'-0" 5'-0"
SHEET No 1

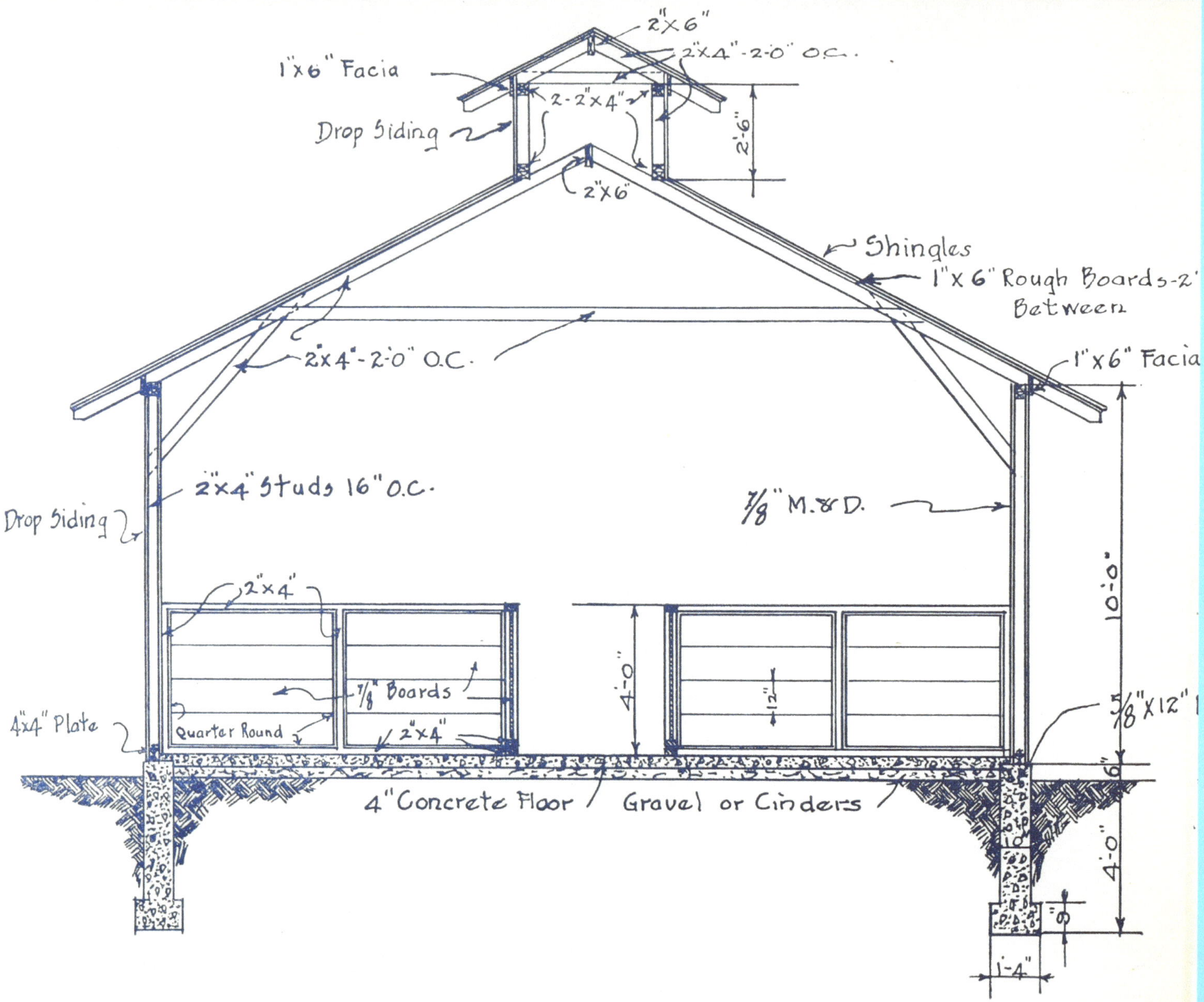

SECTION ON LINE "B-B"

WHS150123, one image, split page

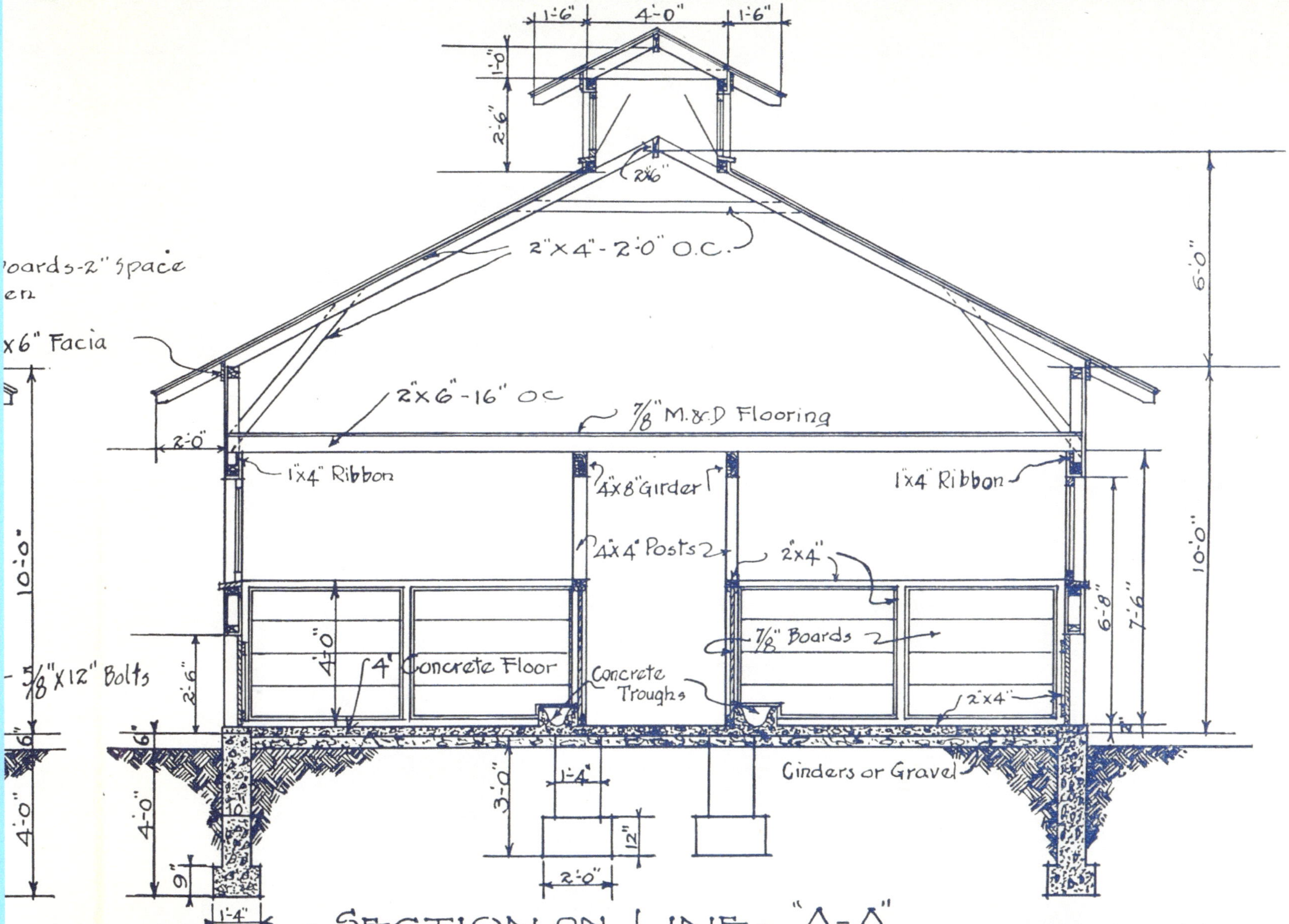

SECTION ON LINE "A-A"

SCALE

PLANS FOR HOG HOUSE

INTERNATIONAL HARVESTER COMPANY OF AMERICA (INCORPORATED)

SHEET No 3.

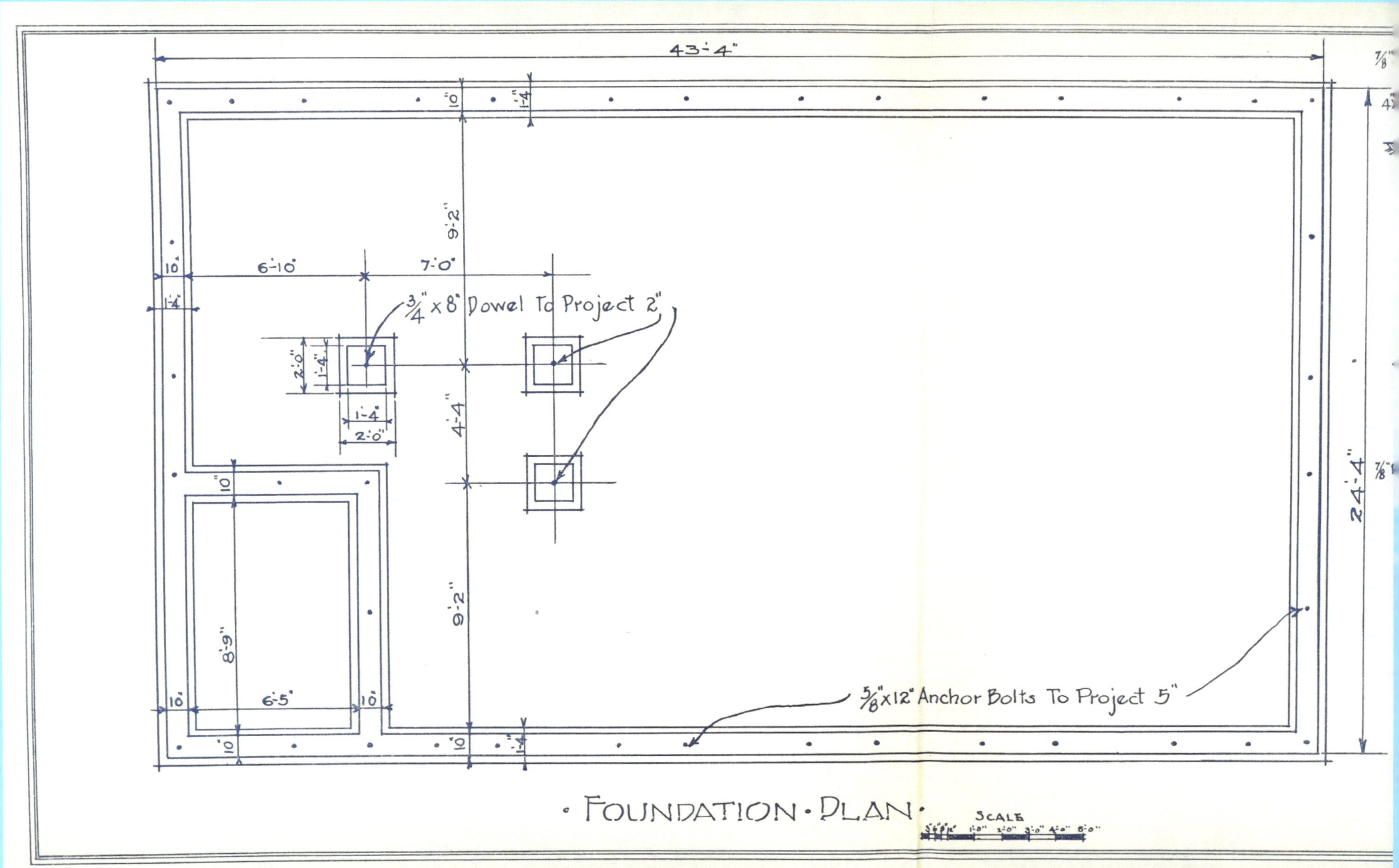
43'-4"
24'-4"
9'-2"
4'-4"
9'-2"
10"
1'-4"
6'-10"
7'-0"
3/4" x 8" Dowel To Project 2"
2'-0"
1'-4"
8'-9"
6'-5"
5/8" x 12" Anchor Bolts To Project 5"
• FOUNDATION • PLAN •
SCALE

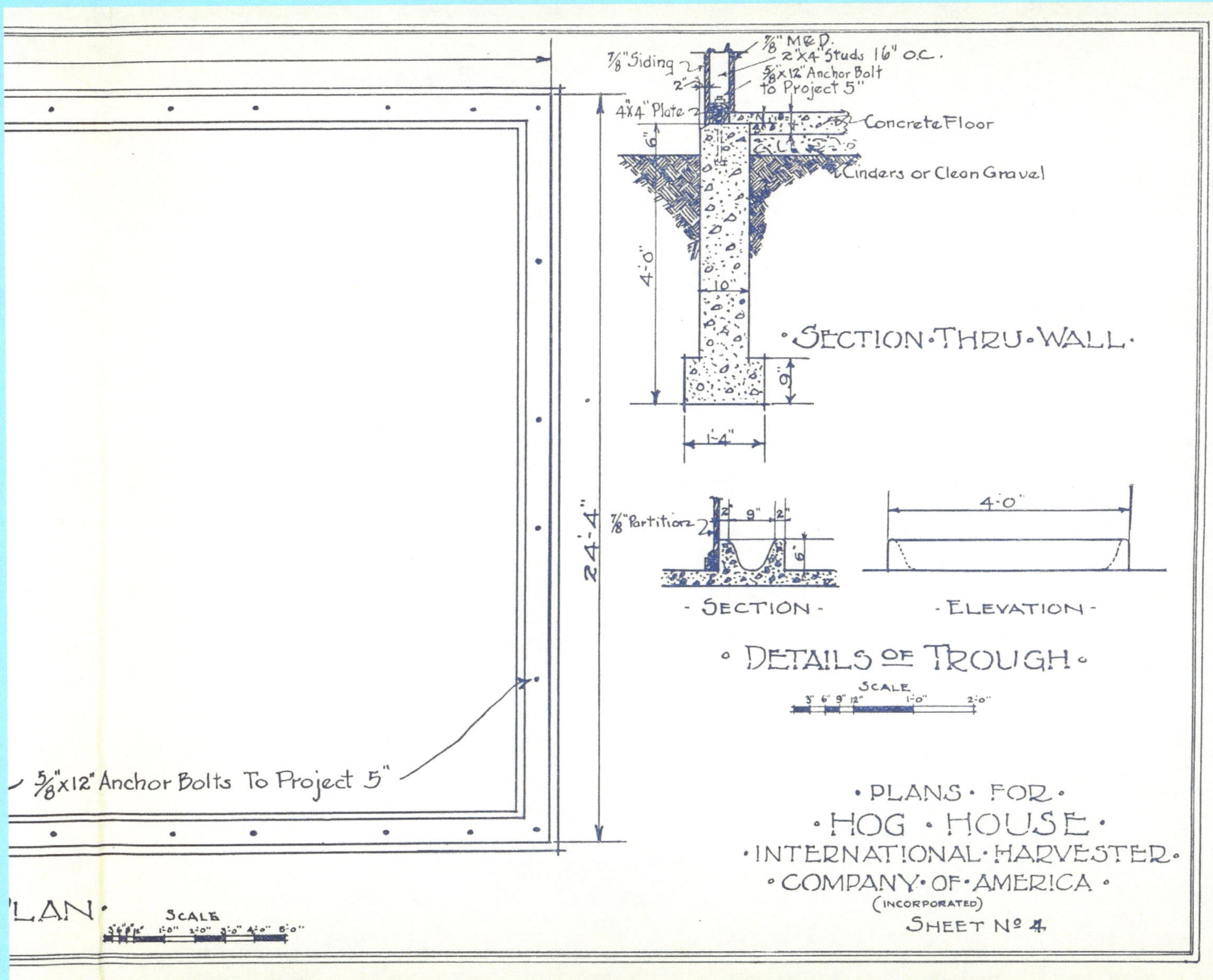

WHS150124, one image, expanded for views.

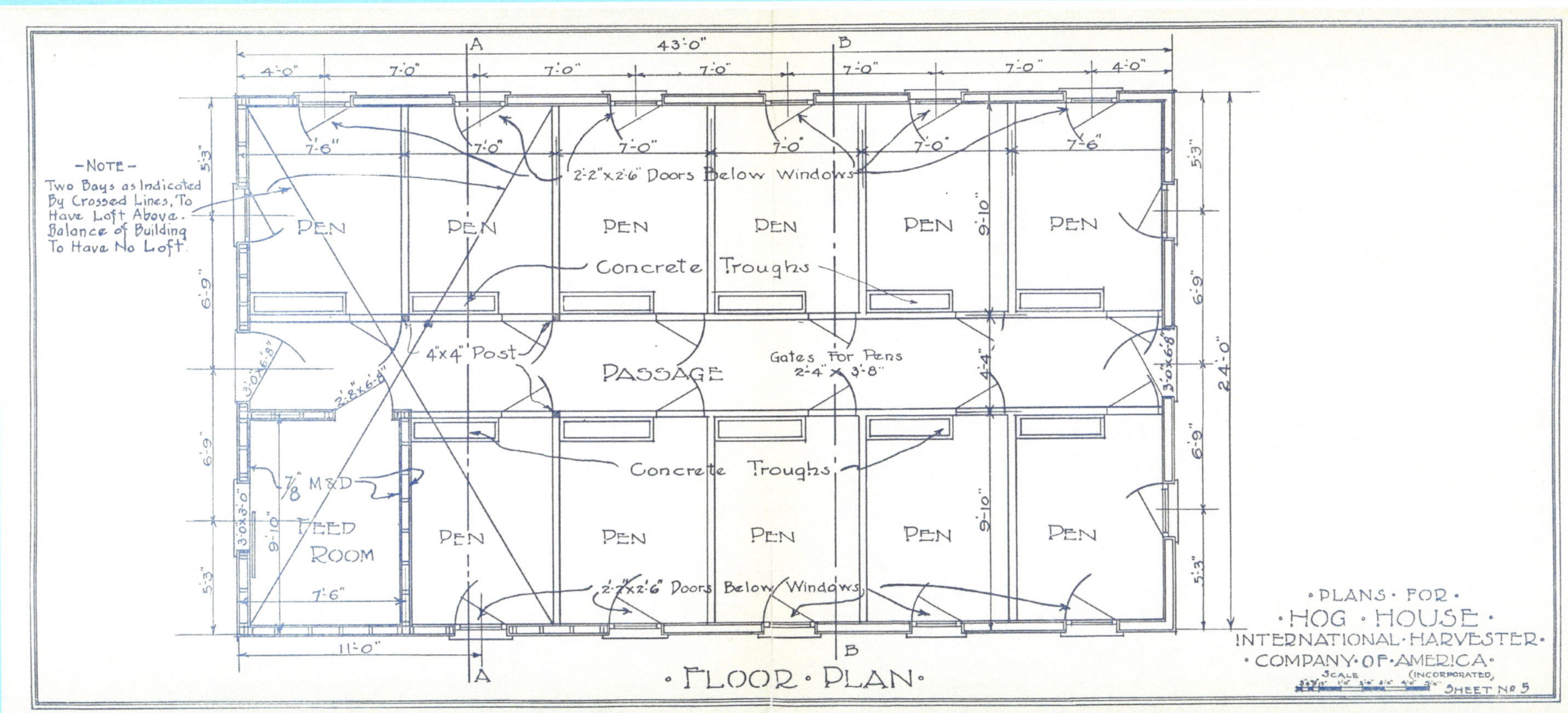
-NOTE-
Two Bays as Indicated By Crossed Lines, To Have Loft Above. Balance of Building To Have No Loft.
43'-0"
PEN
2'-2"x2'-6" Doors Below Windows
Concrete Troughs
4"x4" Post
PASSAGE
Gates For Pens 2'-4" x 3'-8"
7/8" M&D
FEED ROOM
24'-0"
· FLOOR · PLAN ·
· PLANS · FOR ·
· HOG · HOUSE ·
INTERNATIONAL · HARVESTER ·
· COMPANY · OF · AMERICA ·
(INCORPORATED)
SHEET No 5

WHS150125

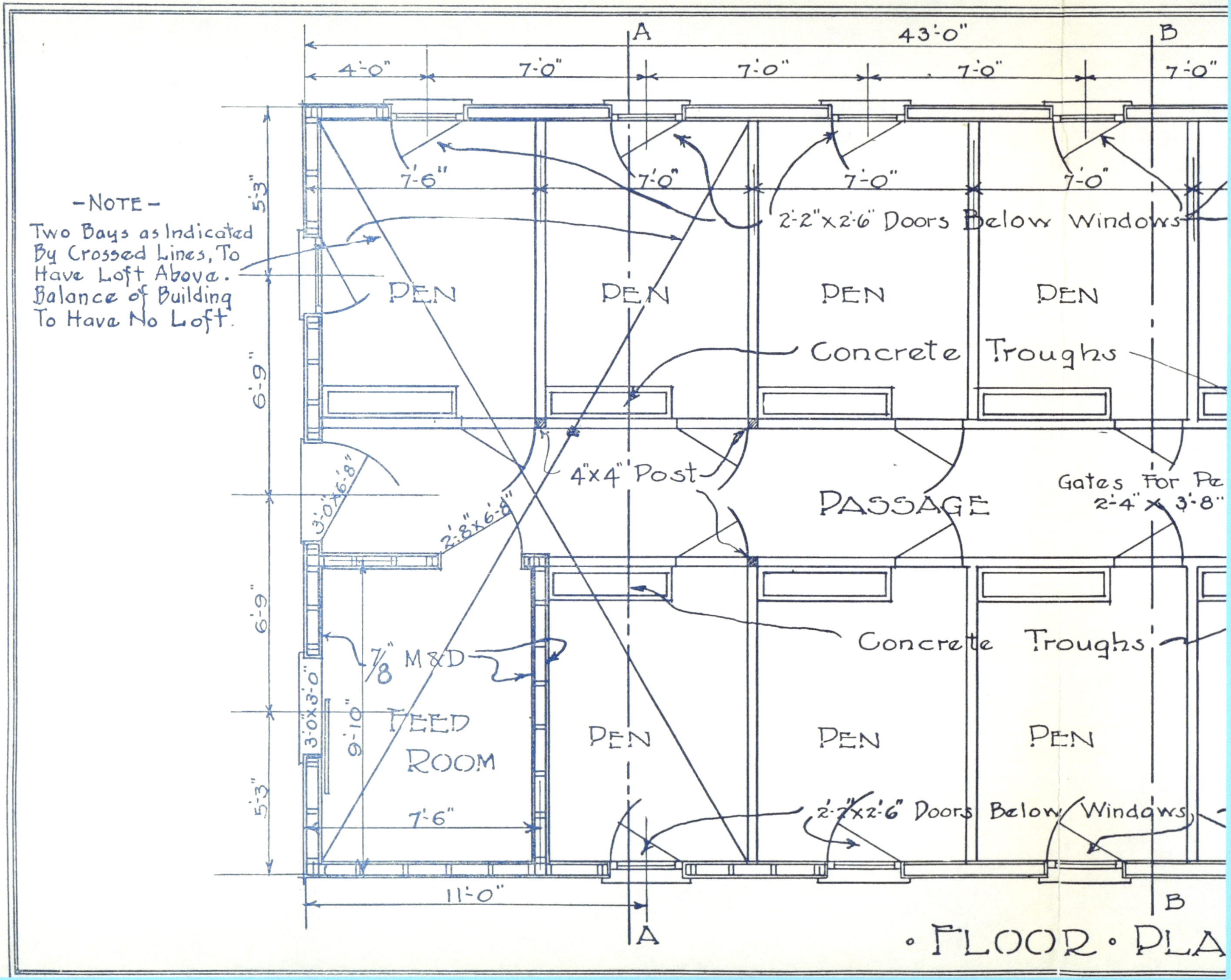

43'-0"
4'-0"
7'-0"
7'-0"
7'-0"
7'-0"
7'-6"
7'-0"
7'-0"
7'-0"
5'-3"
6'-9"
6'-9"
5'-3"
- NOTE -
Two Bays as Indicated By Crossed Lines, To Have Loft Above. Balance of Building To Have No Loft.
2'-2"x2'-6" Doors Below Windows
PEN
PEN
PEN
PEN
Concrete Troughs
4"x4" Post
PASSAGE
Gates For Pe
2'-4" x 3'-8"
3'-0"x6'-8"
2'-8"x6'-8"
7/8" M&D
Concrete Troughs
3'-0"x3'-0"
9'-10"
FEED ROOM
PEN
PEN
PEN
7'-6"
2'-2"x2'-6" Doors Below Windows
11'-0"
A
A
B
B
• FLOOR • PLA

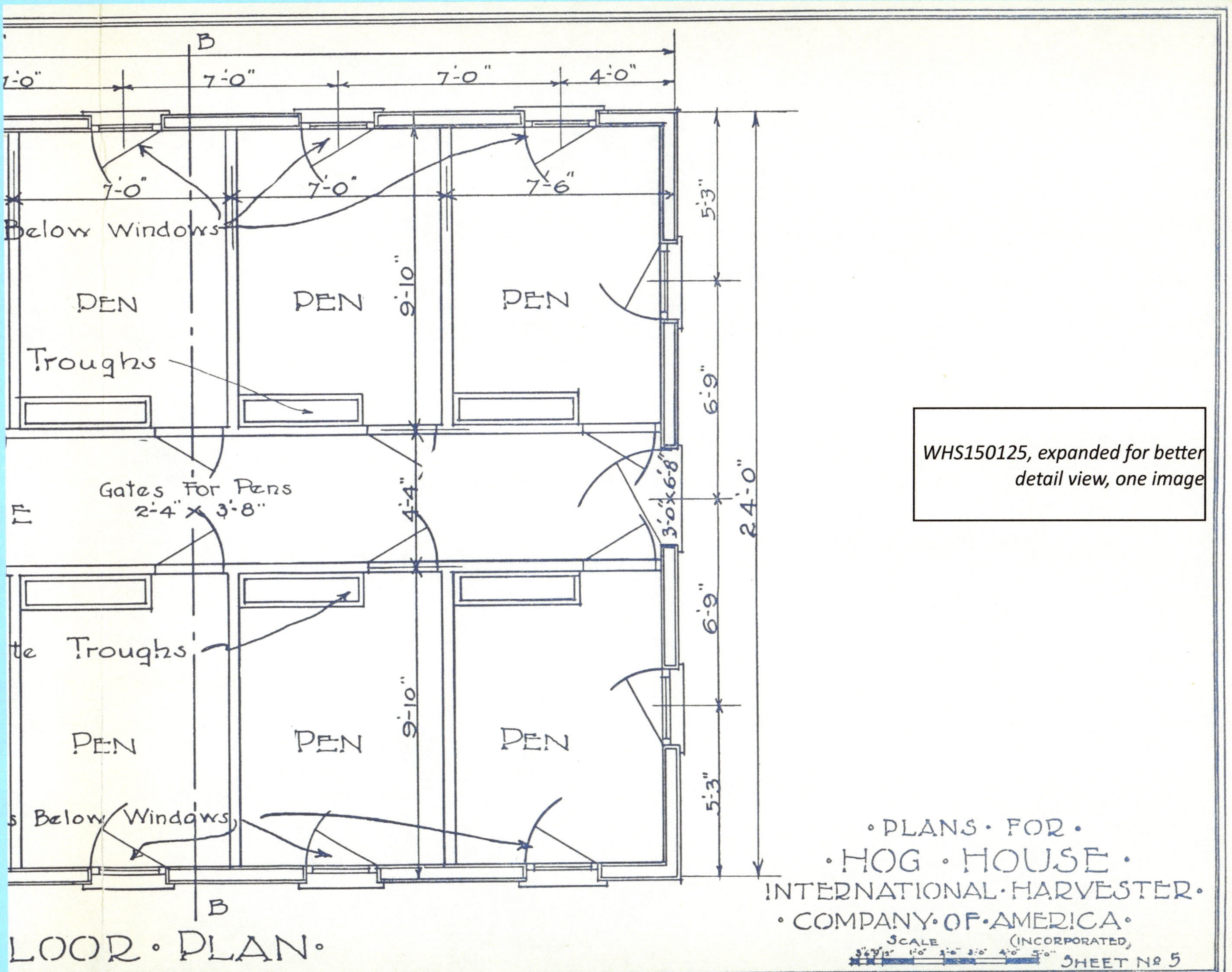

WHS150125, expanded for better detail view, one image

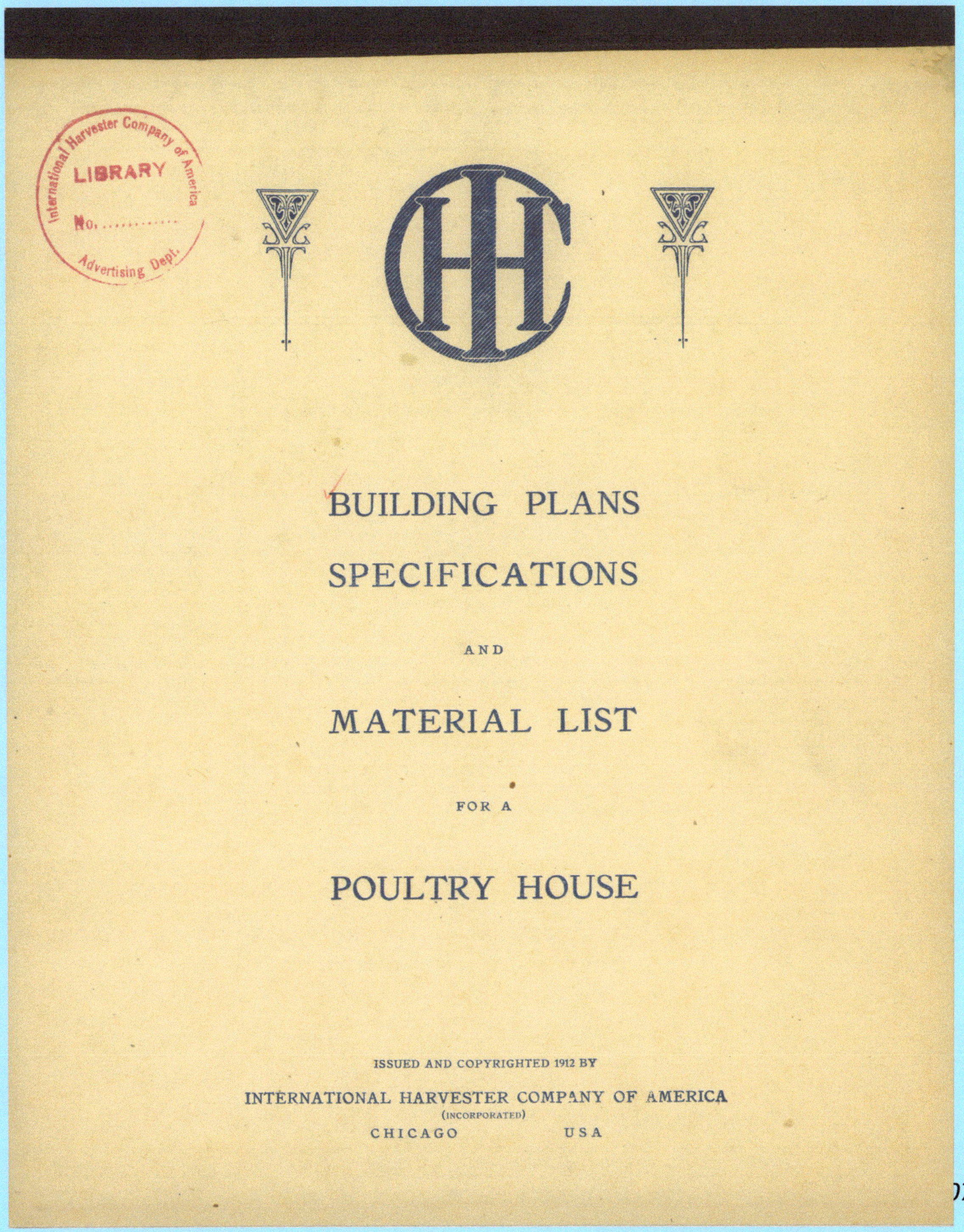

BUILDING PLANS

SPECIFICATIONS

AND

MATERIAL LIST

FOR A

POULTRY HOUSE

ISSUED AND COPYRIGHTED 1912 BY

INTERNATIONAL HARVESTER COMPANY OF AMERICA
(INCORPORATED)
CHICAGO USA

Ms. Carrie Holden, the wife of Perry Holden who was the head of International Harvester Agriculture Extension Service, was a leader in the poultry world and many agricultural colleges modeled their programs from her methods of raising and selling chickens and eggs.

WHS147495

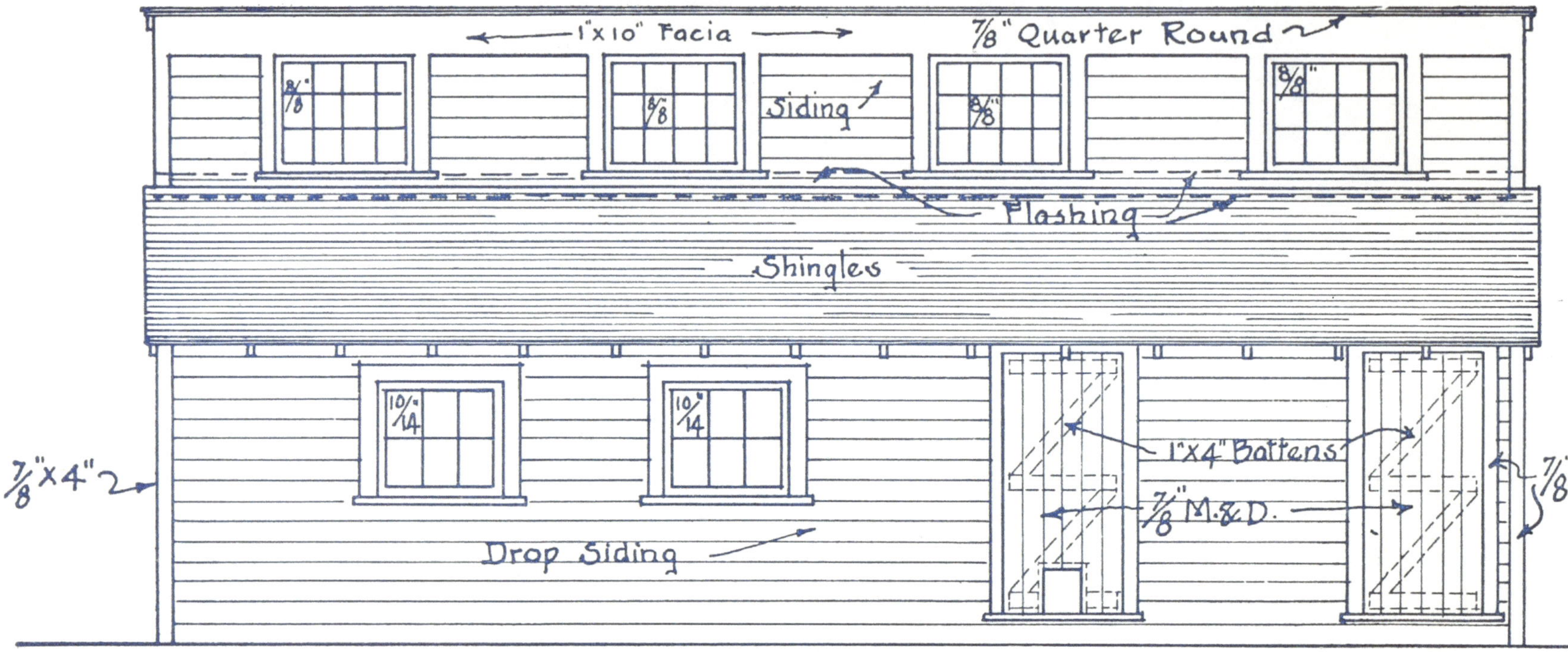

·FRONT· ELEVATION·

WHS150118, one image, split page.

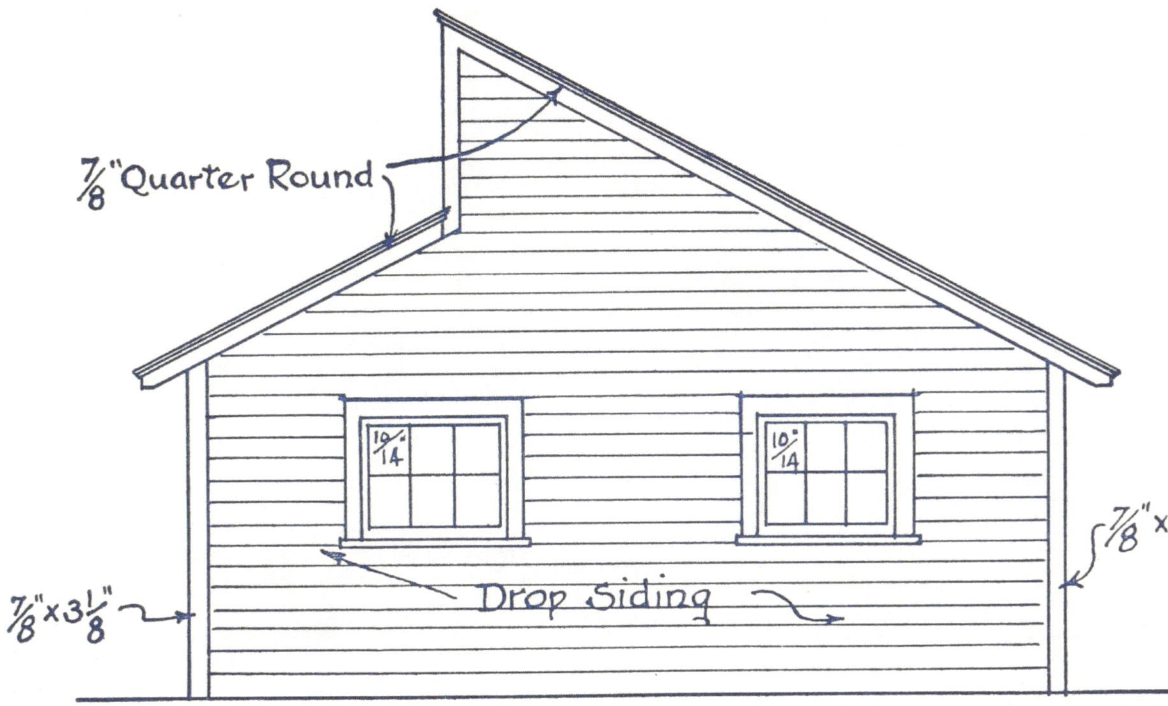

· END · ELEVATION ·

· PLANS · FOR ·

· POULTRY · HOUSE ·

· INTERNATIONAL · HARVESTER ·
· COMPANY · OF · AMERICA ·
(INCORPORATED)

SHEET No. 1

SCALE

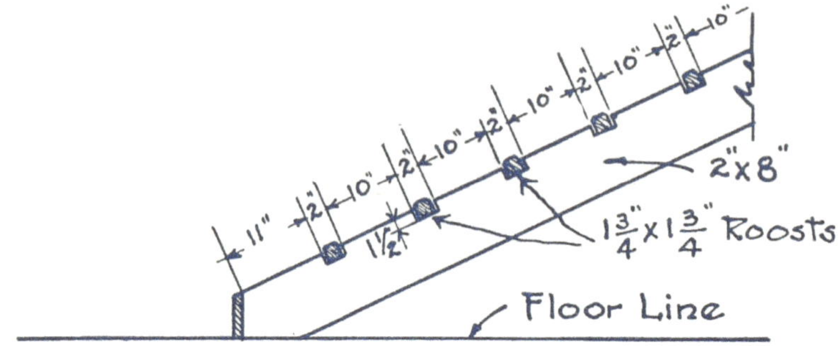

DETAIL OF ROOSTS

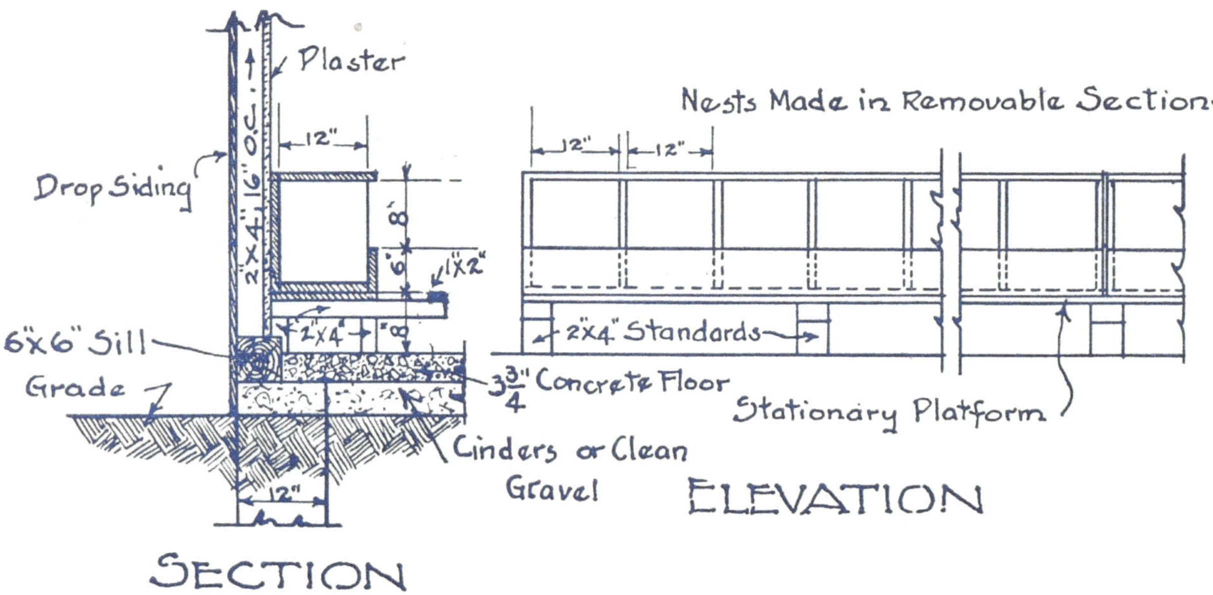

· DETAILS · OF · NESTS ·

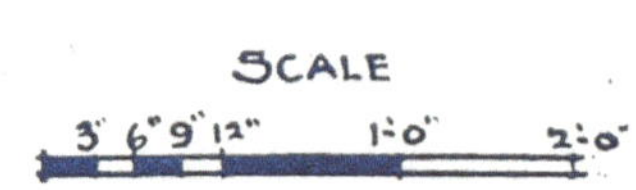

WHS150119, one image, split page

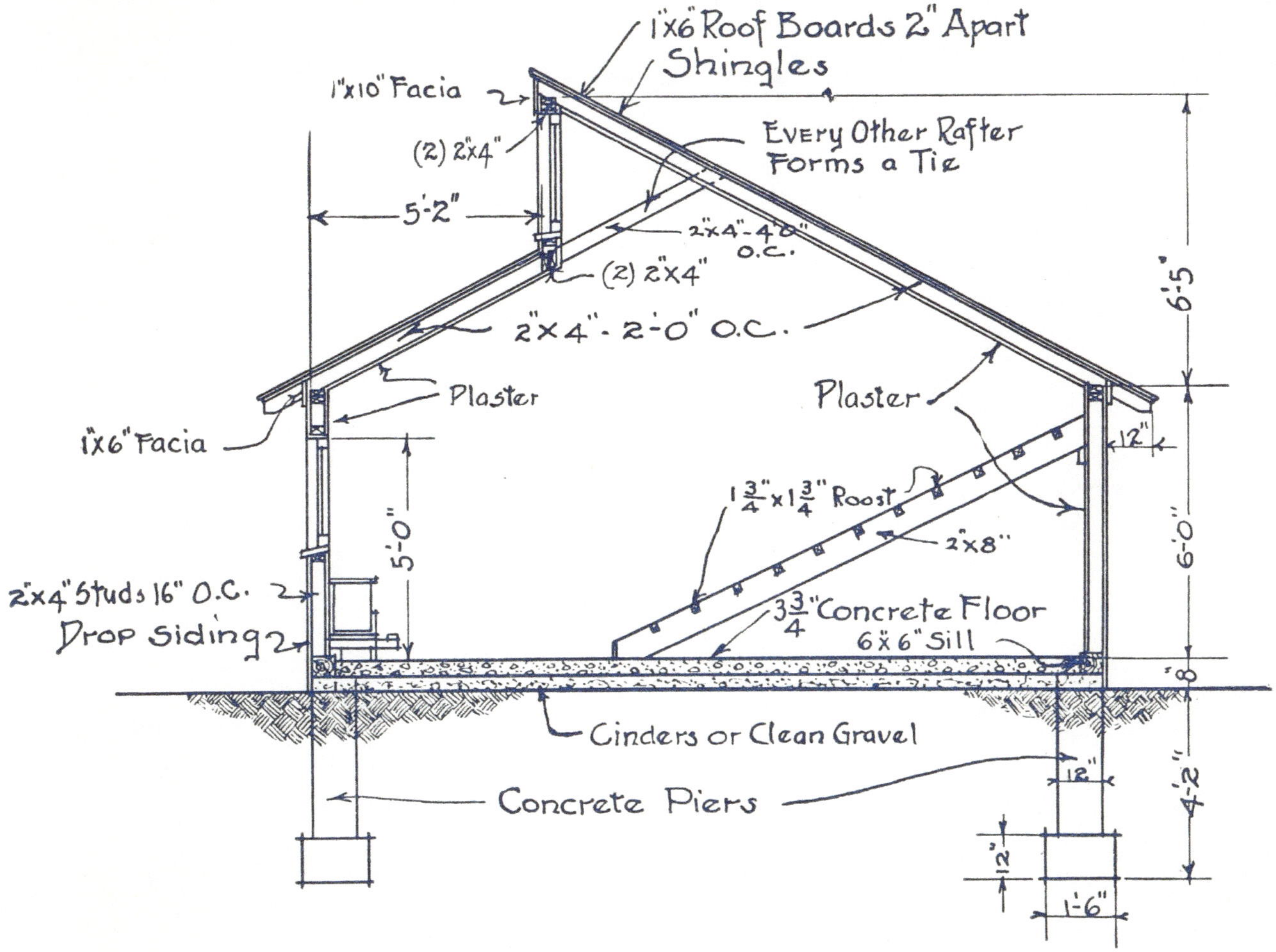

· CROSS · SECTION ·

SCALE

3" 6" 9" 12" 1'-0" 2'-0" 3'-0" 4'-0"

· PLANS · FOR ·
· POULTRY · HOUSE ·
· INTERNATIONAL · HARVESTER ·
· COMPANY · OF · AMERICA ·
(INCORPORATED)

SHEET Nº 2

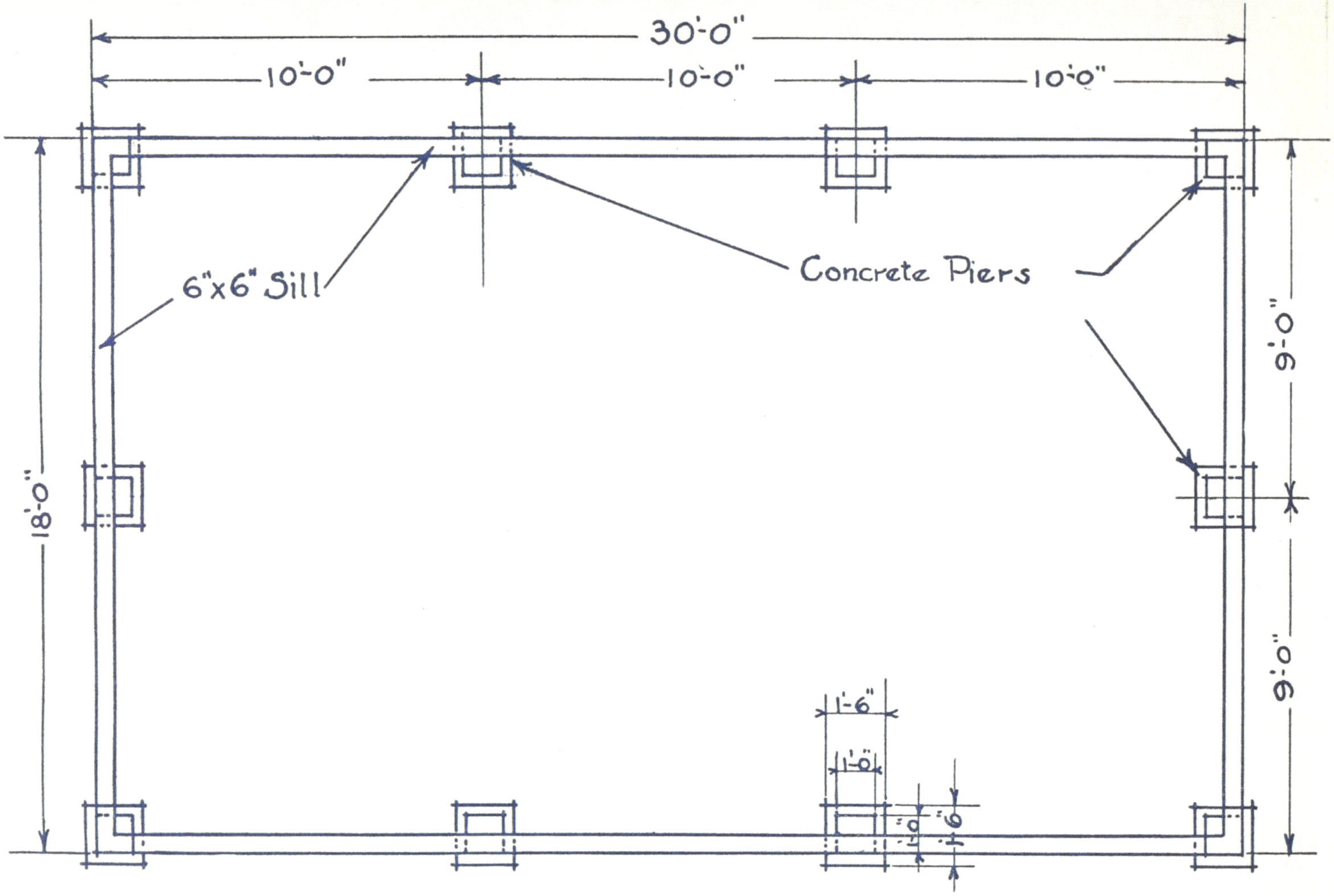

• FOUNDATION • PLAN •

Note- All Wind
Screen

WHS150120, One image, split page

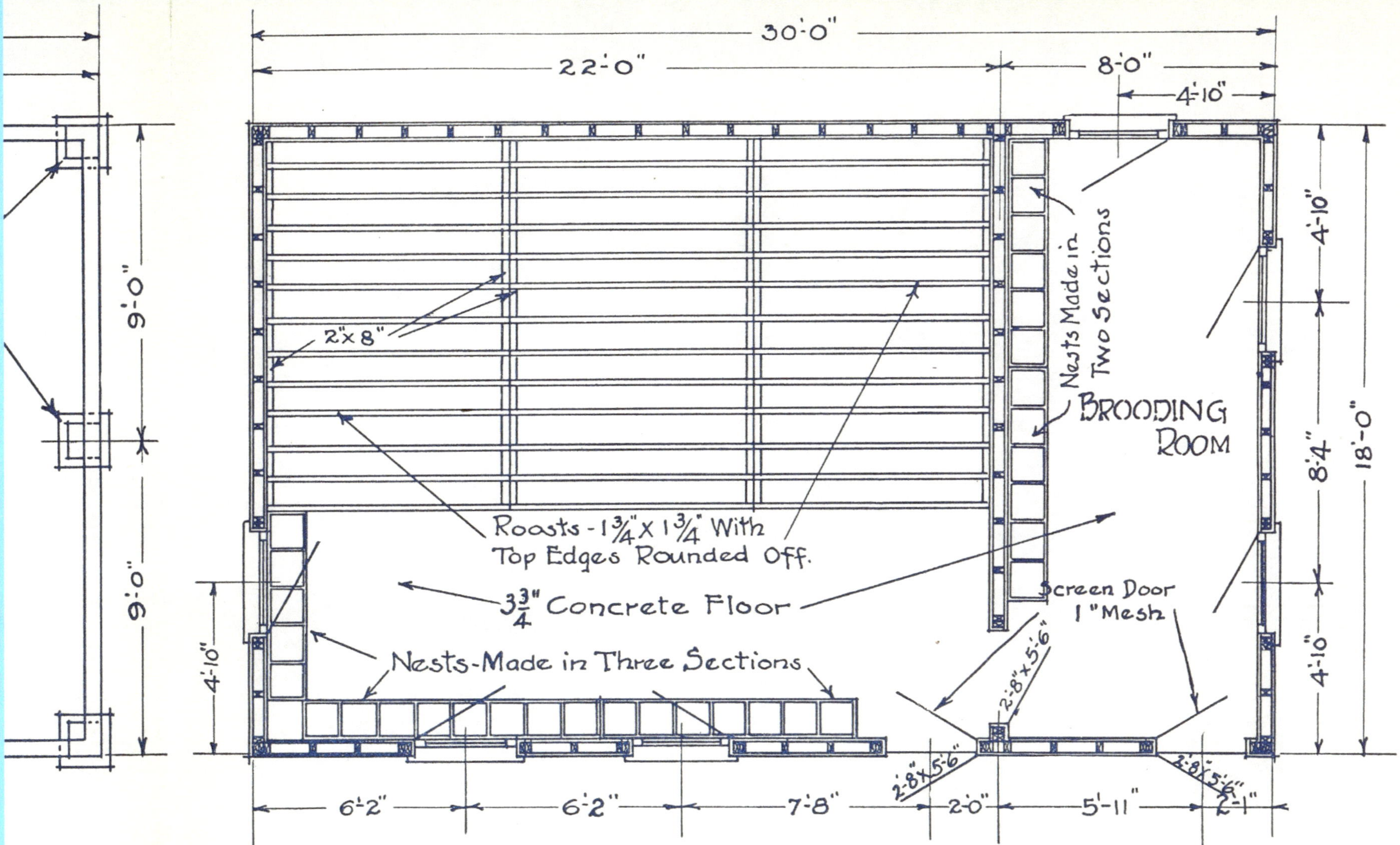

Note- All Windows to be Screened With 1" Mesh

· FLOOR · PLAN ·

· PLANS · FOR ·
· POULTRY · HOUSE ·
· INTERNATIONAL · HARVESTER ·
· COMPANY · OF · AMERICA ·
(INCORPORATED)

SHEET Nº 3

SCALE

Following are plans to expand the Branch House of International Harvester in Syracuse, New York. Expansion Drawings are from November 1935.

Plans are from Tomac Collection.

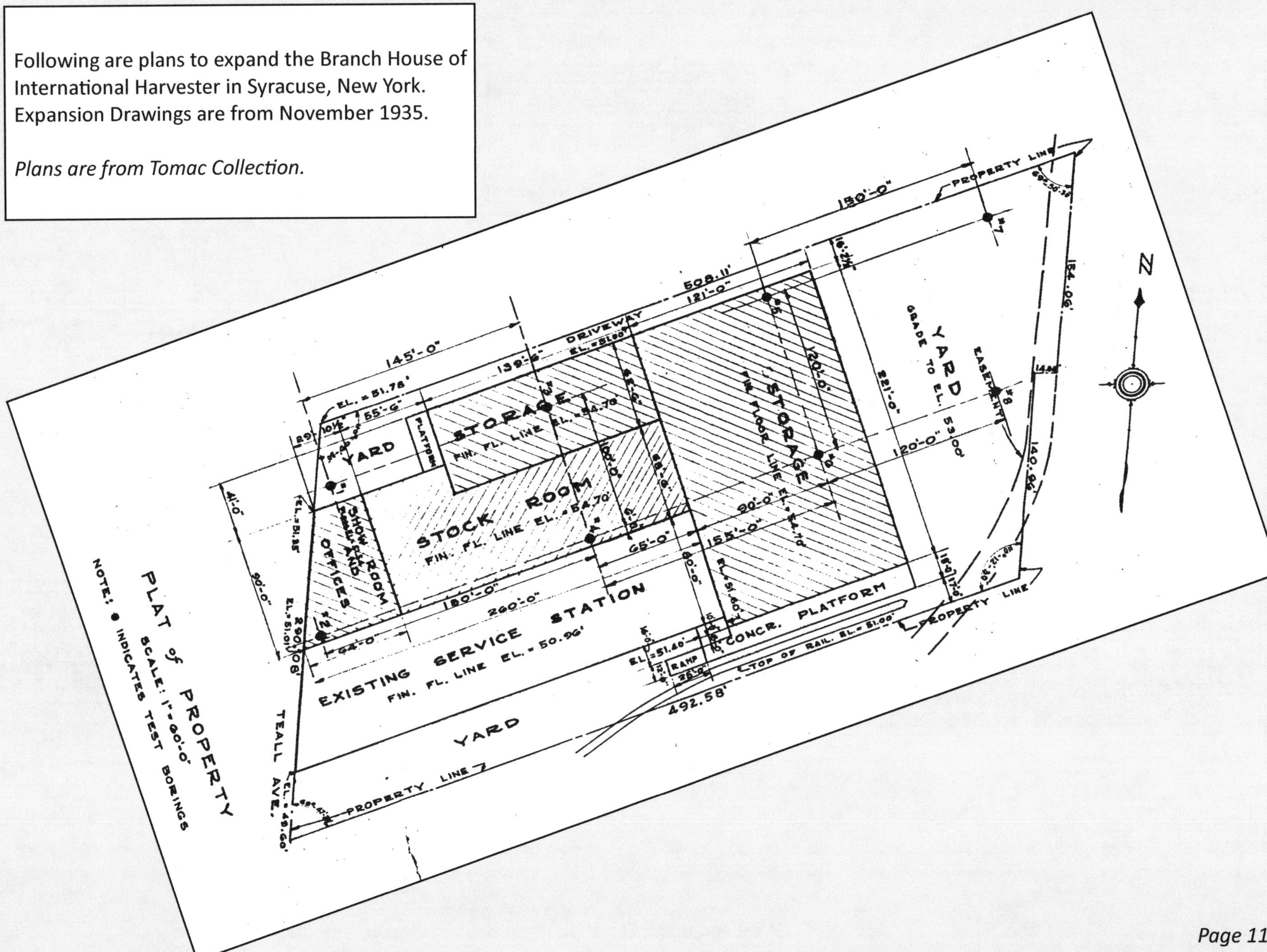

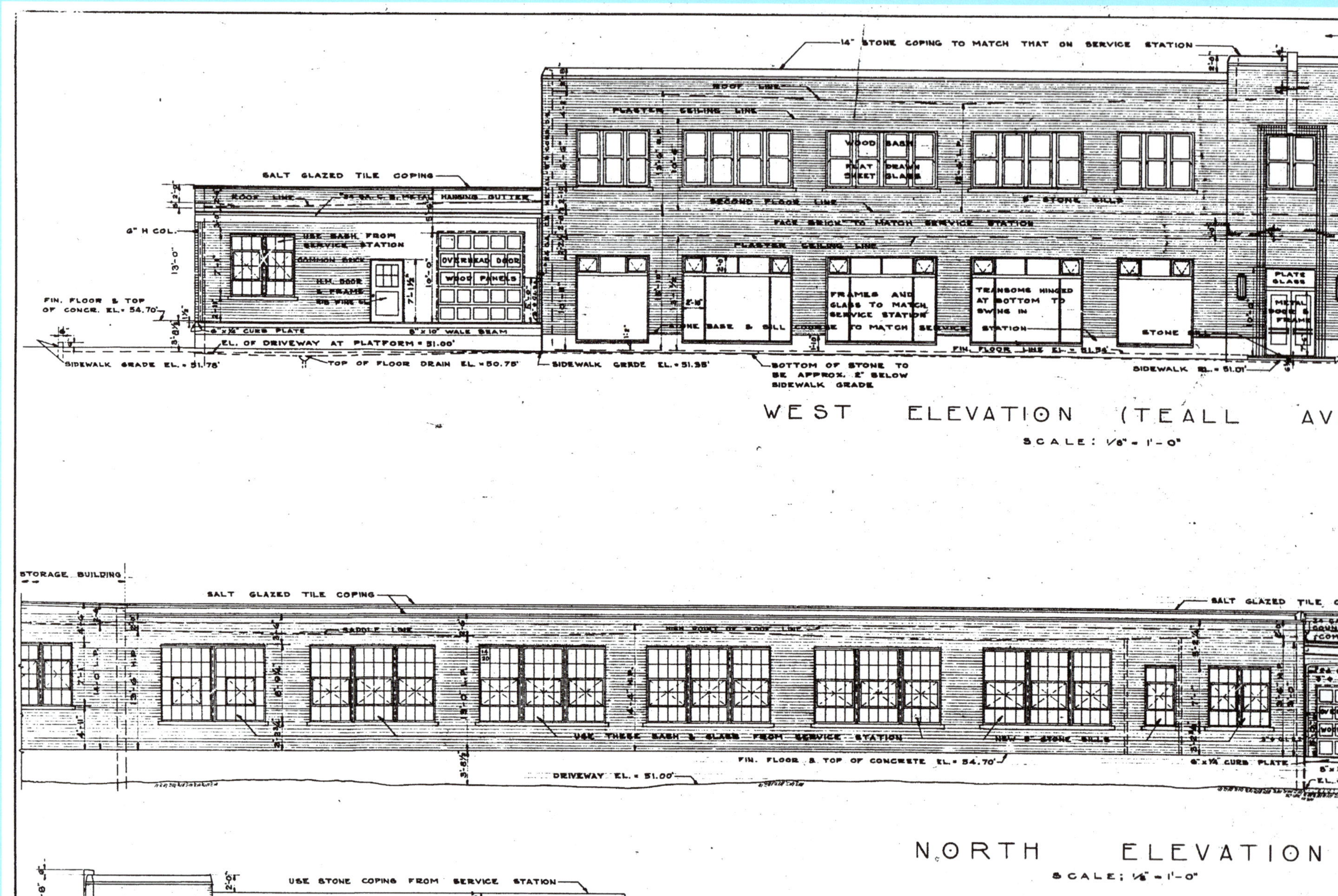

West and North Elevation expanded, one image, split page.

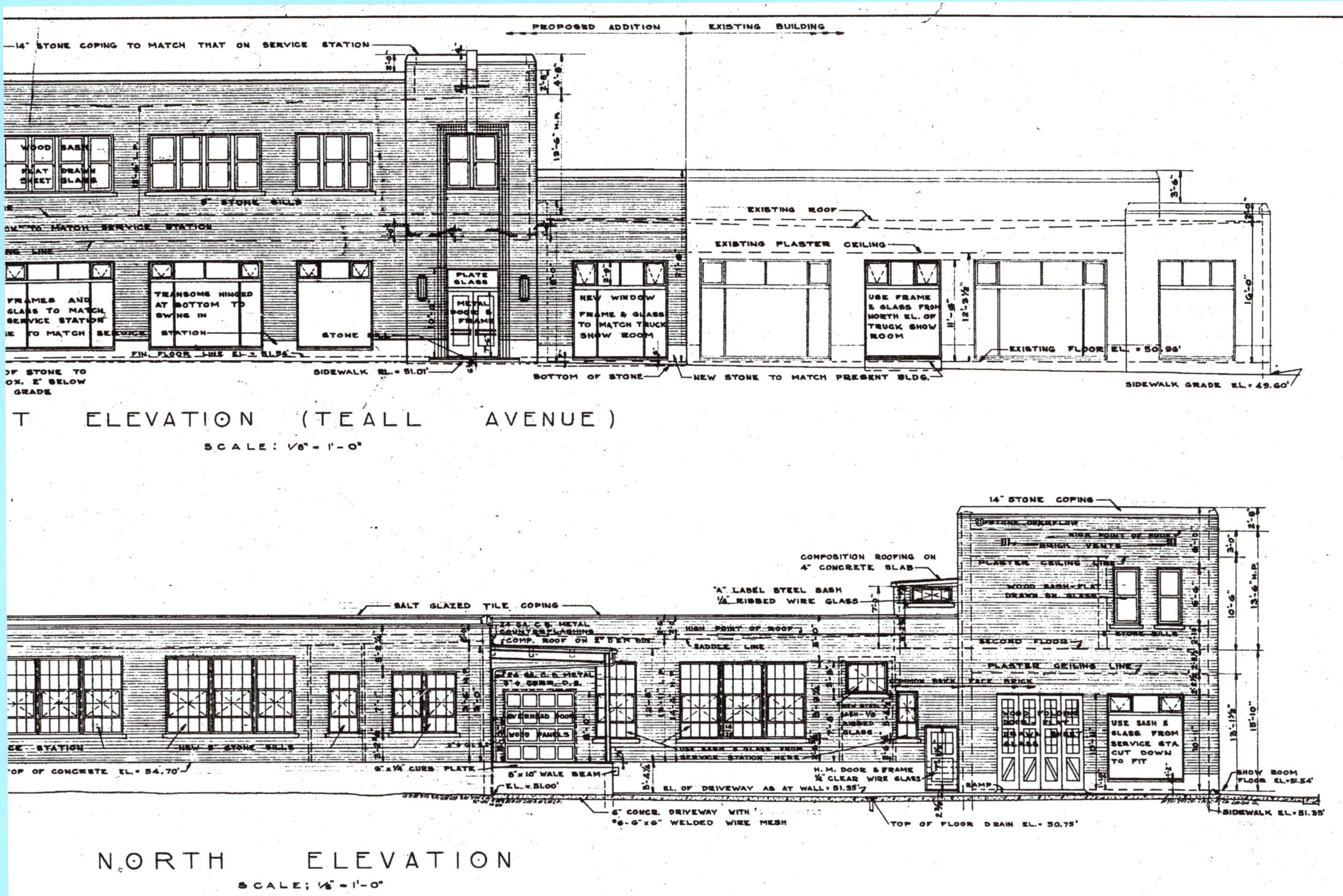

PROPOSED ADDITION
EXISTING BUILDING
14" STONE COPING TO MATCH THAT ON SERVICE STATION
EXISTING ROOF
EXISTING PLASTER CEILING
NEW WINDOW FRAME & GLASS TO MATCH TRUCK SHOW ROOM
USE FRAME & GLASS FROM NORTH EL. OF TRUCK SHOW ROOM
EXISTING FLOOR EL. = 50.96'
SIDEWALK EL. = 51.01'
BOTTOM OF STONE
NEW STONE TO MATCH PRESENT BLDG.
SIDEWALK GRADE EL. = 49.60'
T ELEVATION (TEALL AVENUE)
SCALE: 1/8" = 1'-0"
14" STONE COPING
COMPOSITION ROOFING ON 4" CONCRETE SLAB
"A" LABEL STEEL SASH 1/4 RIBBED WIRE GLASS
SALT GLAZED TILE COPING
SECOND FLOOR
PLASTER CEILING LINE
USE SASH & GLASS FROM SERVICE STA CUT DOWN TO FIT
H.M. DOOR & FRAME 1/4 CLEAR WIRE GLASS
EL. OF DRIVEWAY AS AT WALL = 51.35'
6" CONC. DRIVEWAY WITH 6-6" x 6" WELDED WIRE MESH
TOP OF FLOOR DRAIN EL. = 50.75'
SHOW ROOM FLOOR EL. = 51.54'
SIDEWALK EL. = 51.35'
NORTH ELEVATION
SCALE; 1/8" = 1'-0"

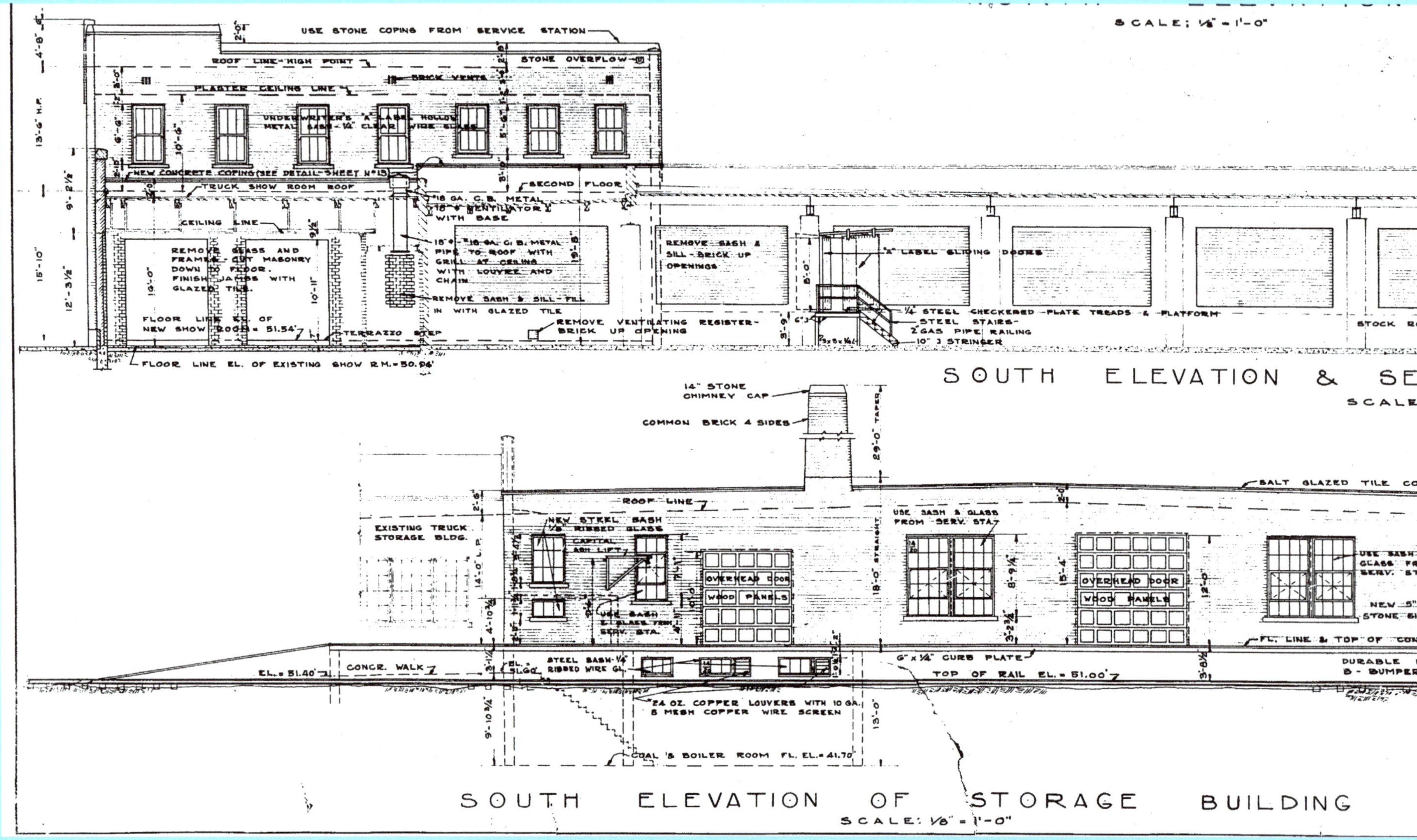

South Elevation expanded, one image, split page, some overlap of images for clarification of images

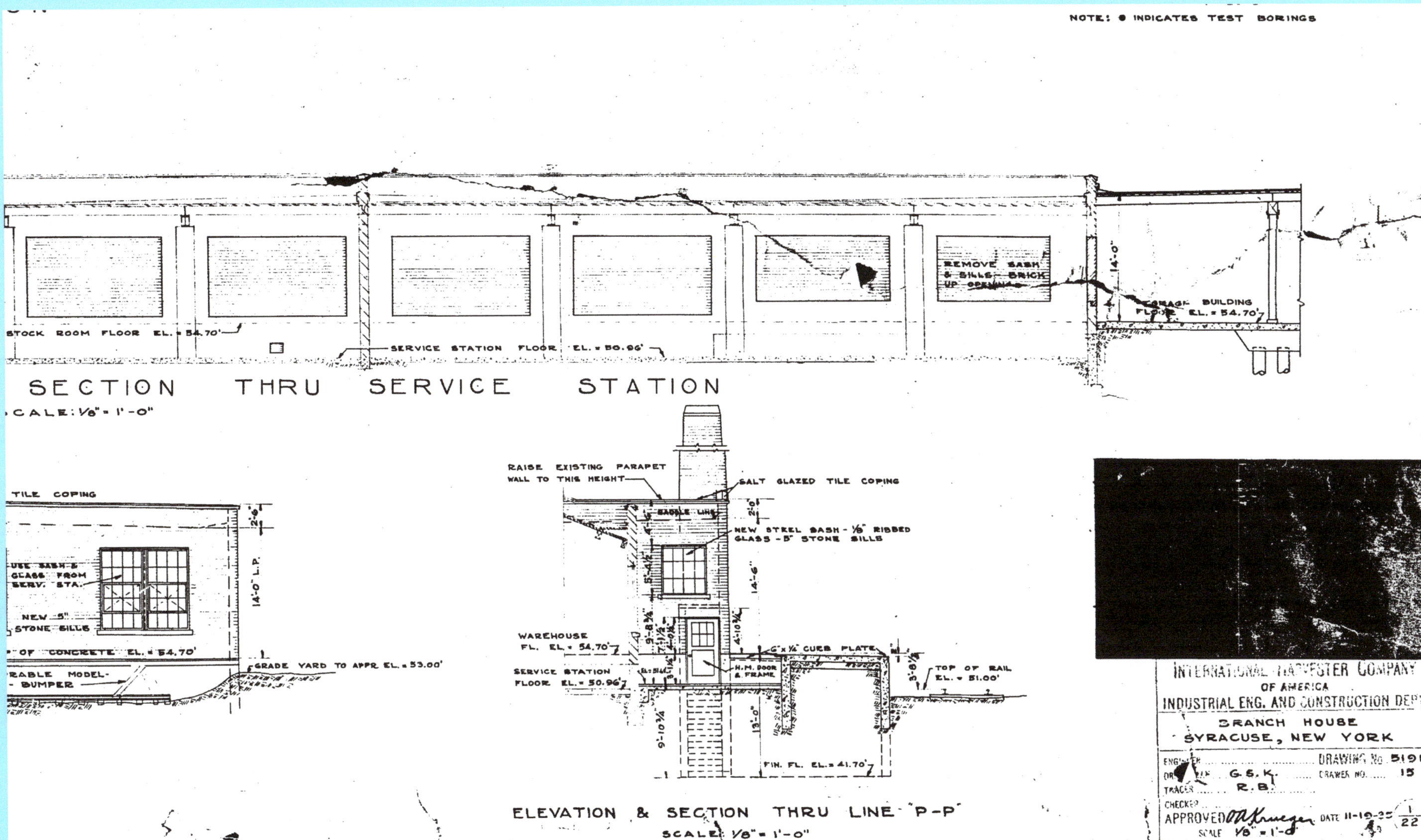
NOTE: ● INDICATES TEST BORINGS
REMOVE SASH & SILLS - BRICK UP OPENING
GARAGE BUILDING FLOOR EL. = 54.70'
14'-0"
STOCK ROOM FLOOR EL. = 54.70'
SERVICE STATION FLOOR EL. = 50.96'
SECTION THRU SERVICE STATION
SCALE: 1/8" = 1'-0"
TILE COPING
USE SASH & GLASS FROM SERV. STA.
NEW 5" STONE SILLS
14'-0" L.P.
2'-6"
OF CONCRETE EL. = 54.70'
GRADE YARD TO APPR. EL. = 53.00'
BUMPER
RAISE EXISTING PARAPET WALL TO THIS HEIGHT
SALT GLAZED TILE COPING
NEW STEEL SASH - 1/8" RIBBED GLASS - 5" STONE SILLS
14'-6"
WAREHOUSE FL. EL. = 54.70'
SERVICE STATION FLOOR EL. = 50.96'
H.M. DOOR & FRAME
6" x 1/4" CURB PLATE
TOP OF RAIL EL. = 51.00'
13'-0"
9'-10 1/4"
FIN. FL. EL. = 41.70'
ELEVATION & SECTION THRU LINE "P-P"
SCALE: 1/8" = 1'-0"
INTERNATIONAL HARVESTER COMPANY OF AMERICA
INDUSTRIAL ENG. AND CONSTRUCTION DEPT.
BRANCH HOUSE
SYRACUSE, NEW YORK
DRAWING No. 5191
DRAWER NO. 15
G. S. K.
R. B.
APPROVED O.A. Krueger
DATE 11-19-25
SCALE 1/8" = 1'-0"
1/22

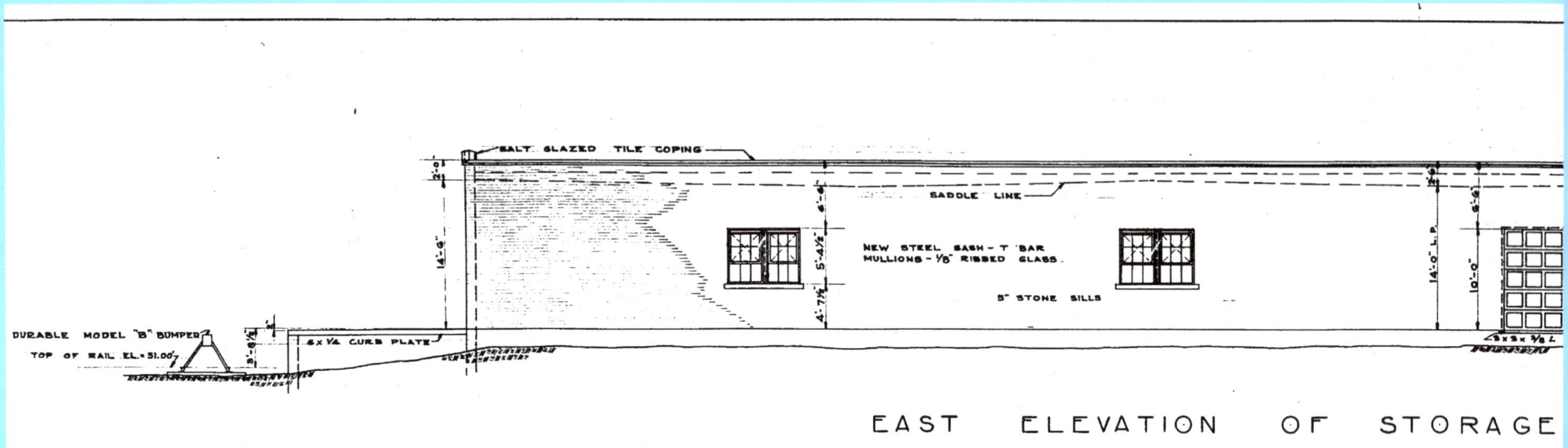

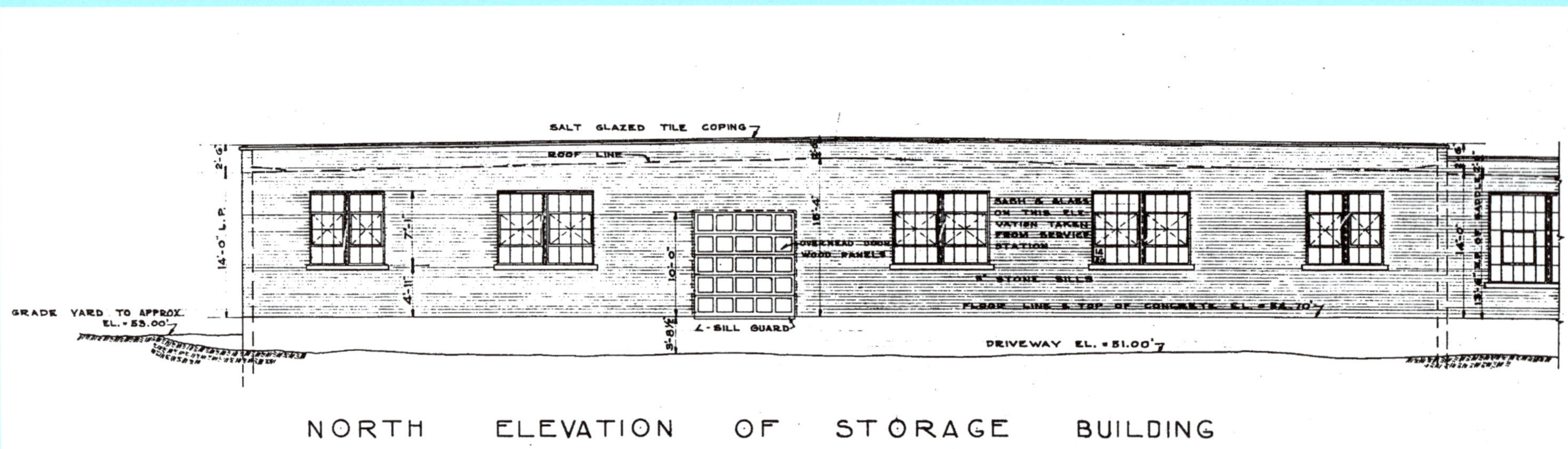

East and North Elevation expanded, one image, split page, some overlap of images for clarification of images

ORAGE BUILDING

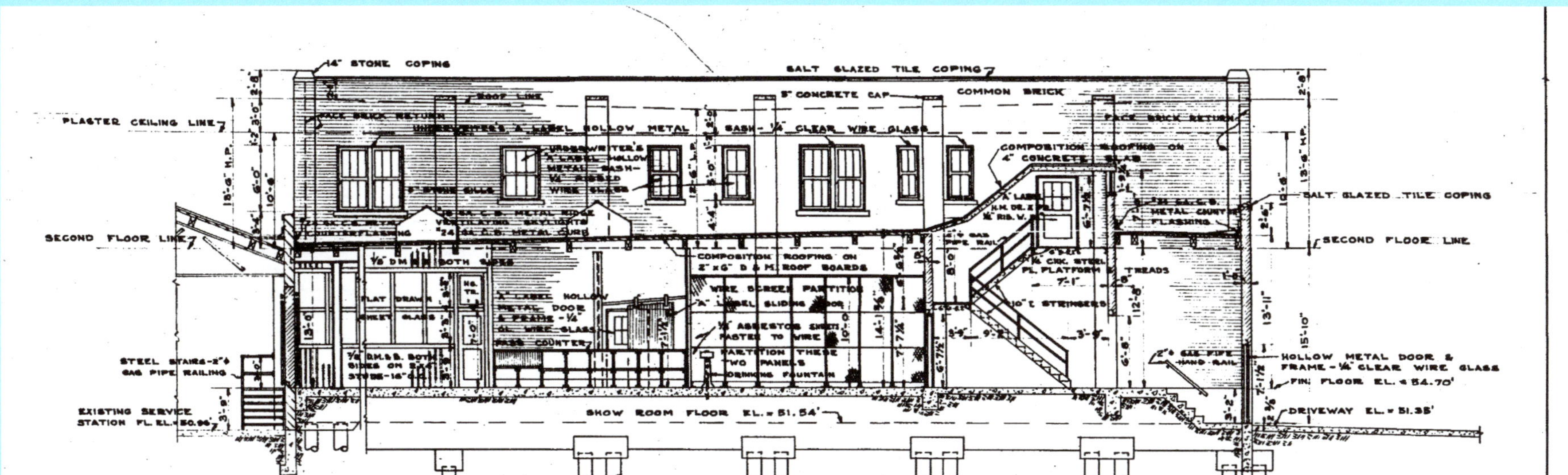

CROSS SECTION AND ELEVATION THRU LINE "A-A"

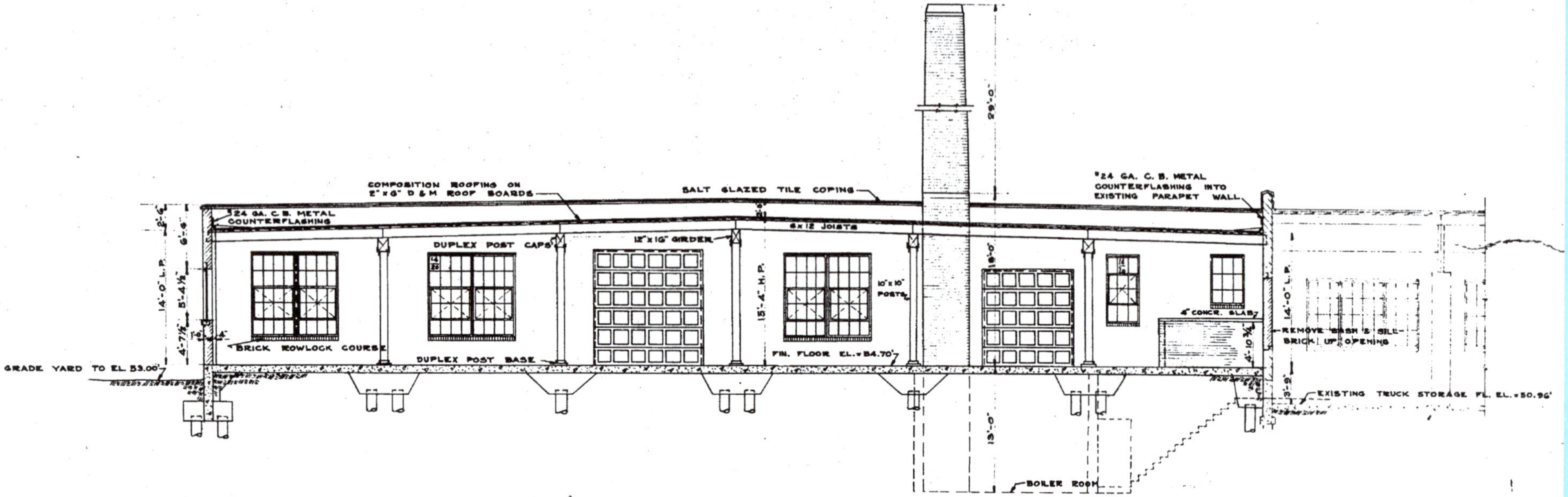

SECTION "N-N" THRU STORAGE BUILDING

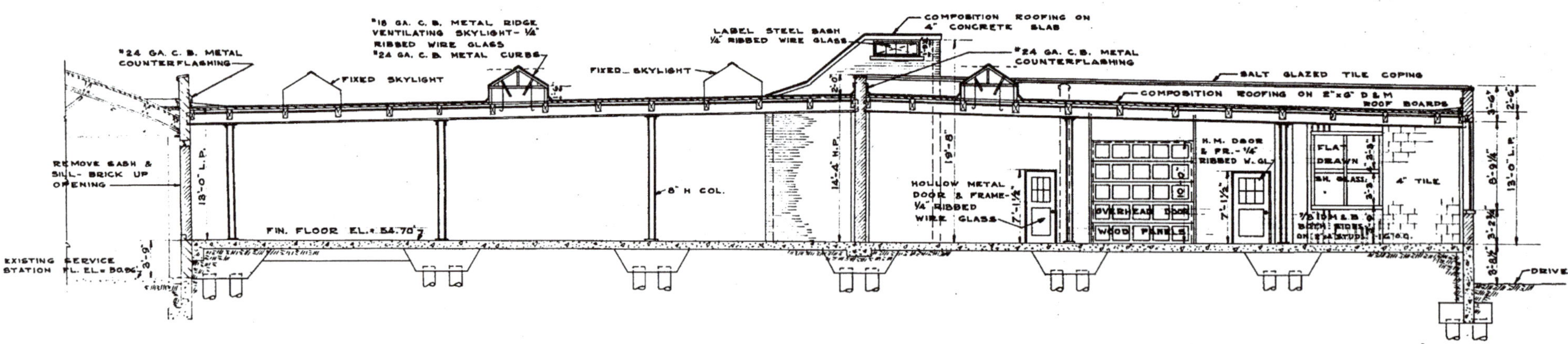

SECTION THRU LINE "B-B"

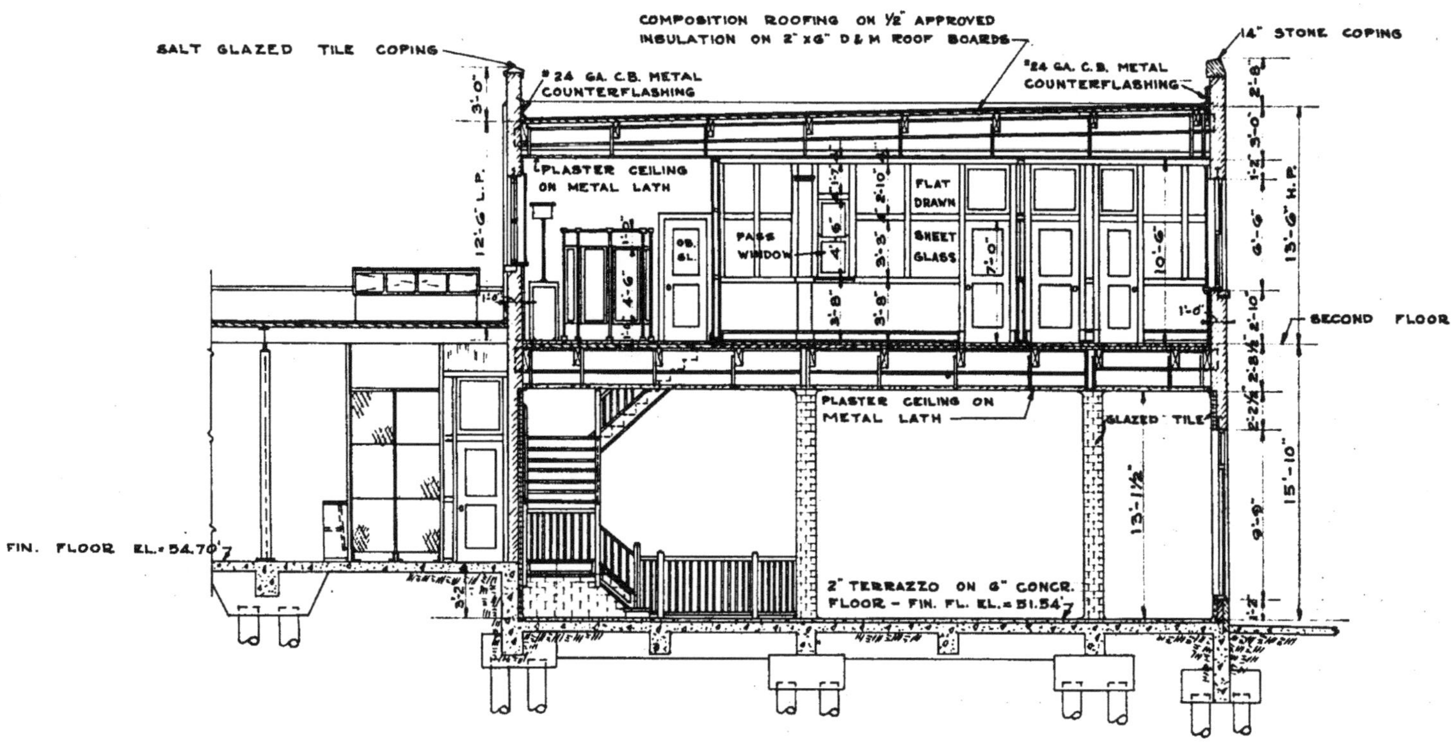

SECTION THRU LINE "C-C"

Thru Sections of the building, service and storage areas.

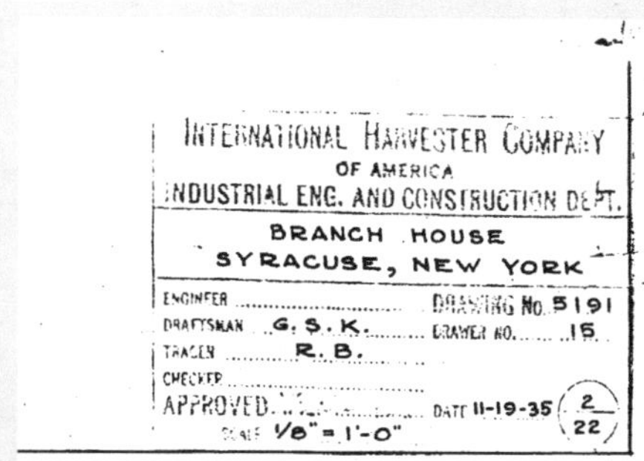

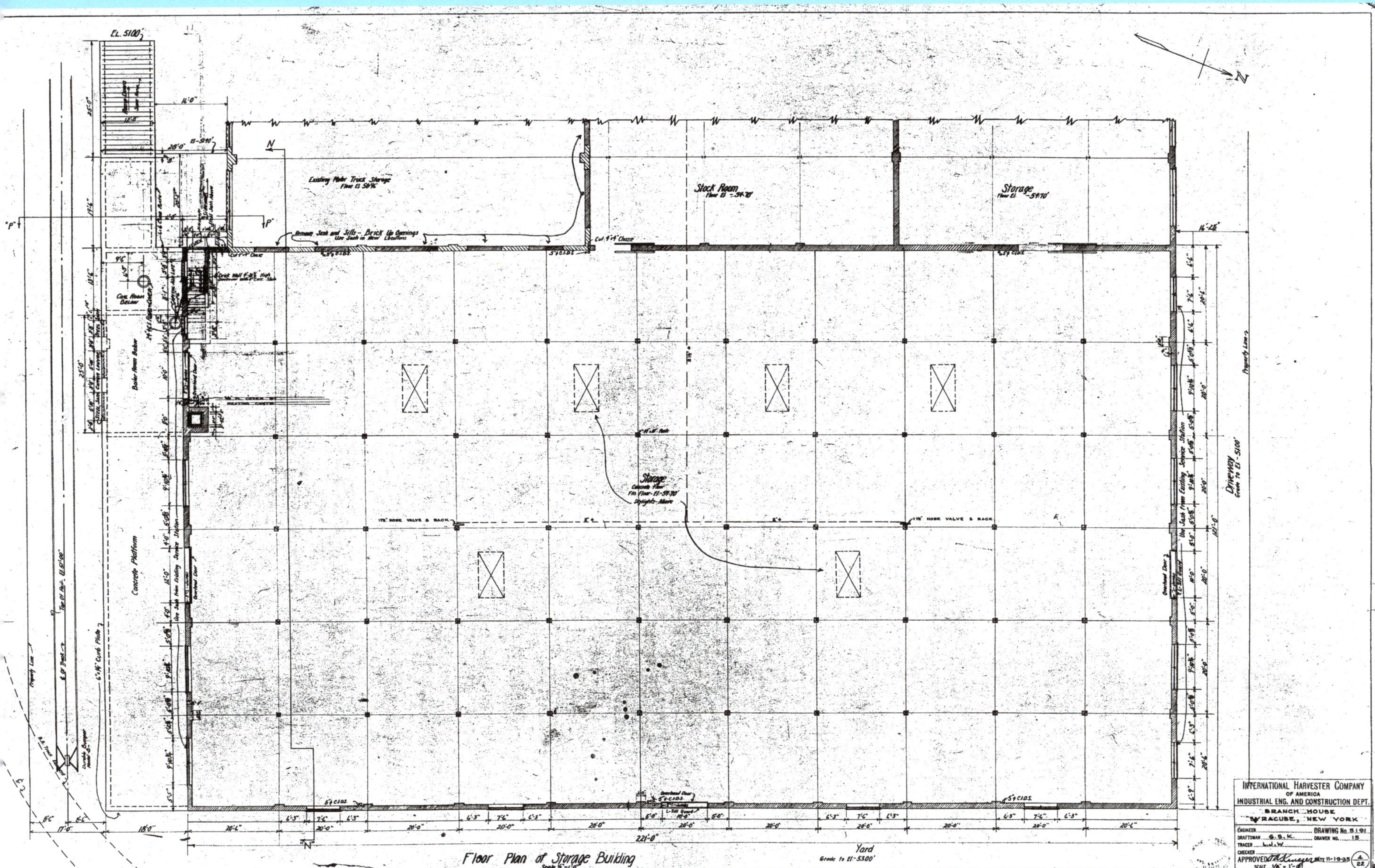

Floor plan of storage building showing skylights.

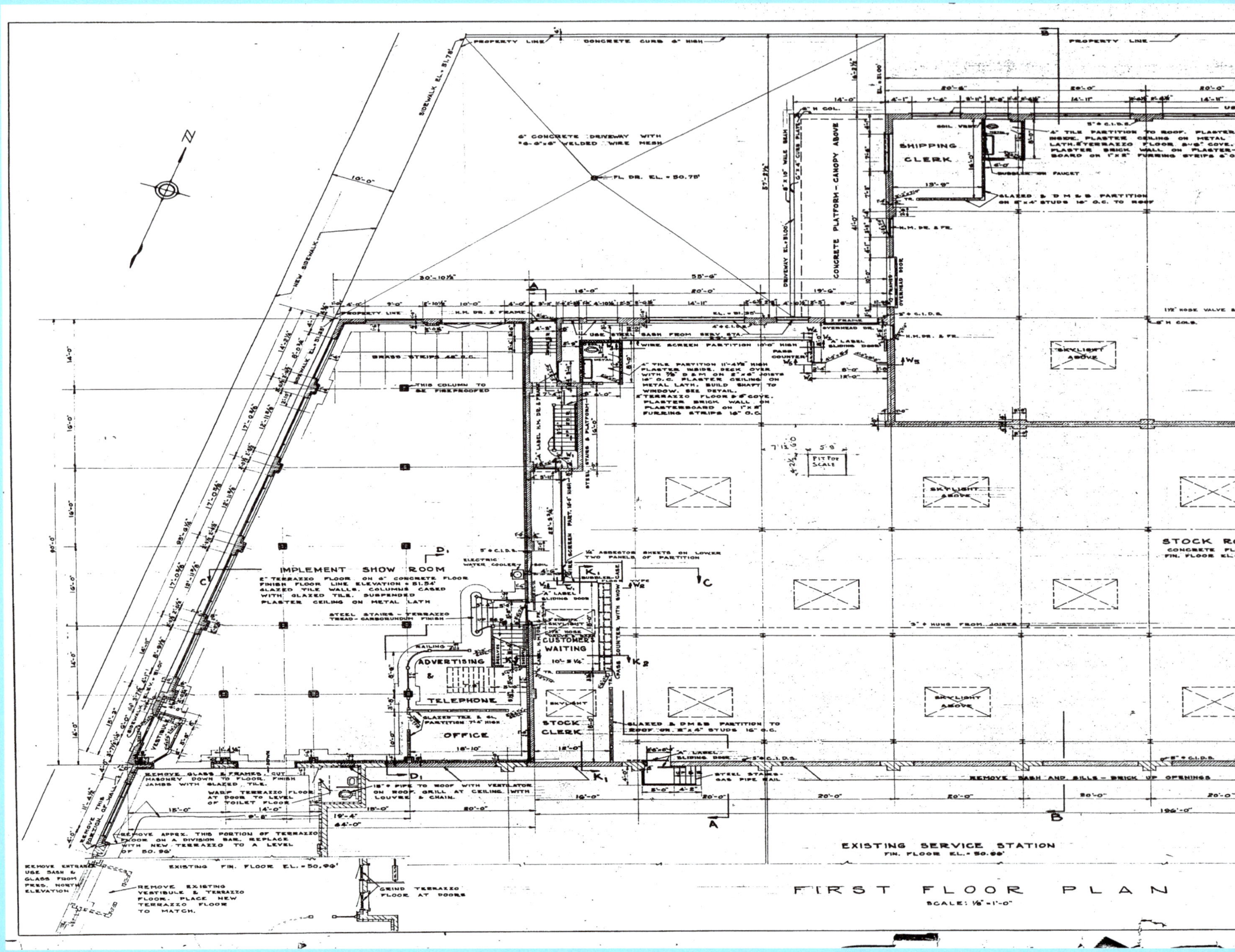

PROPERTY LINE
CONCRETE CURB 6" HIGH
SIDEWALK EL. = 51.78'
NEW SIDEWALK
6" CONCRETE DRIVEWAY WITH 6-6"x6" WELDED WIRE MESH
FL. DR. EL. = 50.75'
CONCRETE PLATFORM - CANOPY ABOVE
SHIPPING CLERK
SKYLIGHT ABOVE
BRASS STRIPS
THIS COLUMN TO BE FIREPROOFED
IMPLEMENT SHOW ROOM
PIT FOR SCALE
STOCK RO
STEEL STAIRS - TERRAZZO TREAD - CARBORUNDUM FINISH
ADVERTISING & TELEPHONE
CUSTOMERS WAITING
OFFICE
STOCK CLERK
REMOVE SASH AND SILLS - BRICK UP OPENINGS
EXISTING SERVICE STATION
EXISTING FIN. FLOOR EL. = 50.99'
FIRST FLOOR PLAN
SCALE: 1/8" = 1'-0"

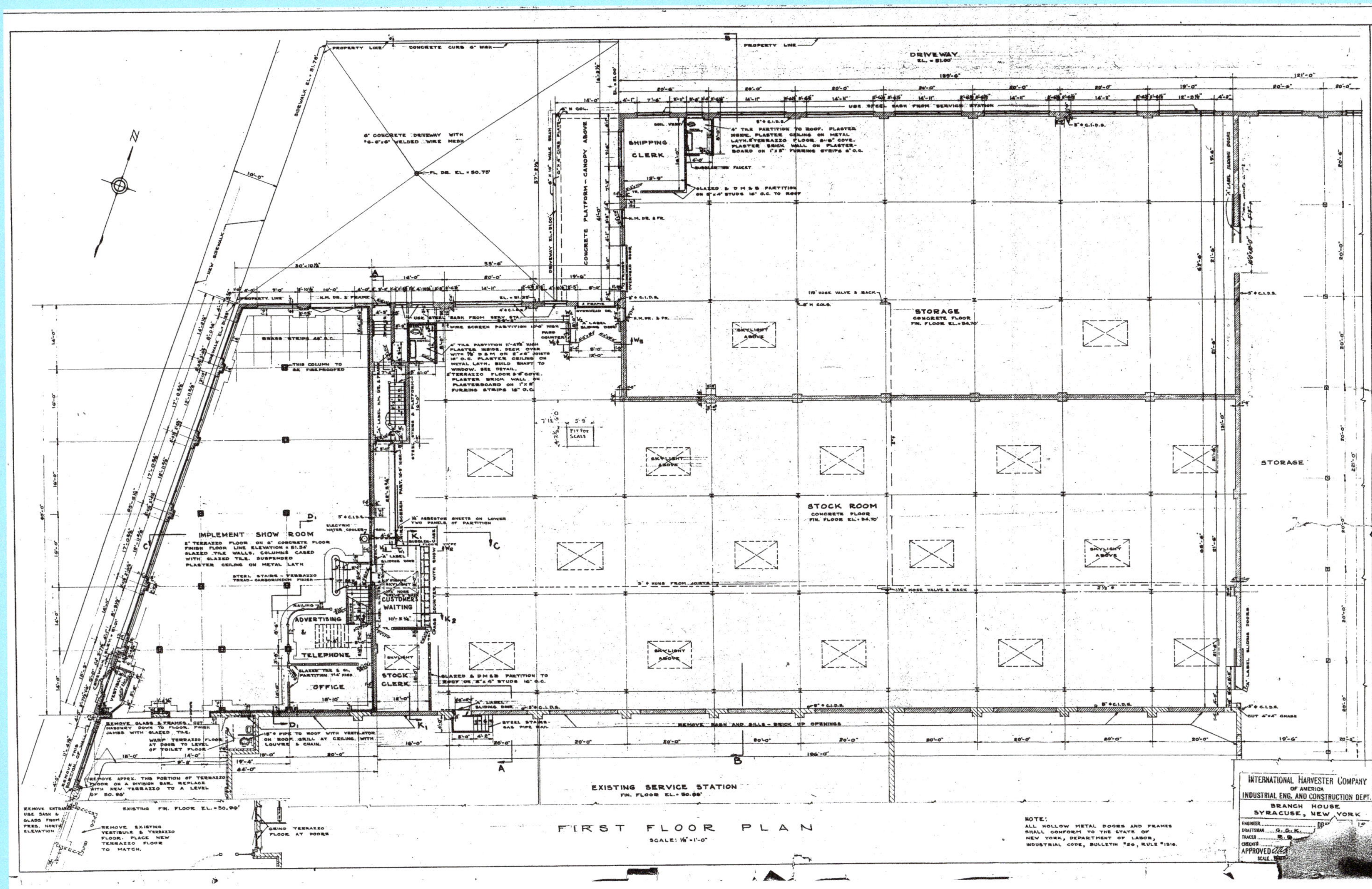

Service station plan, full image on this page, expanded view on previous page for better view of show room area.

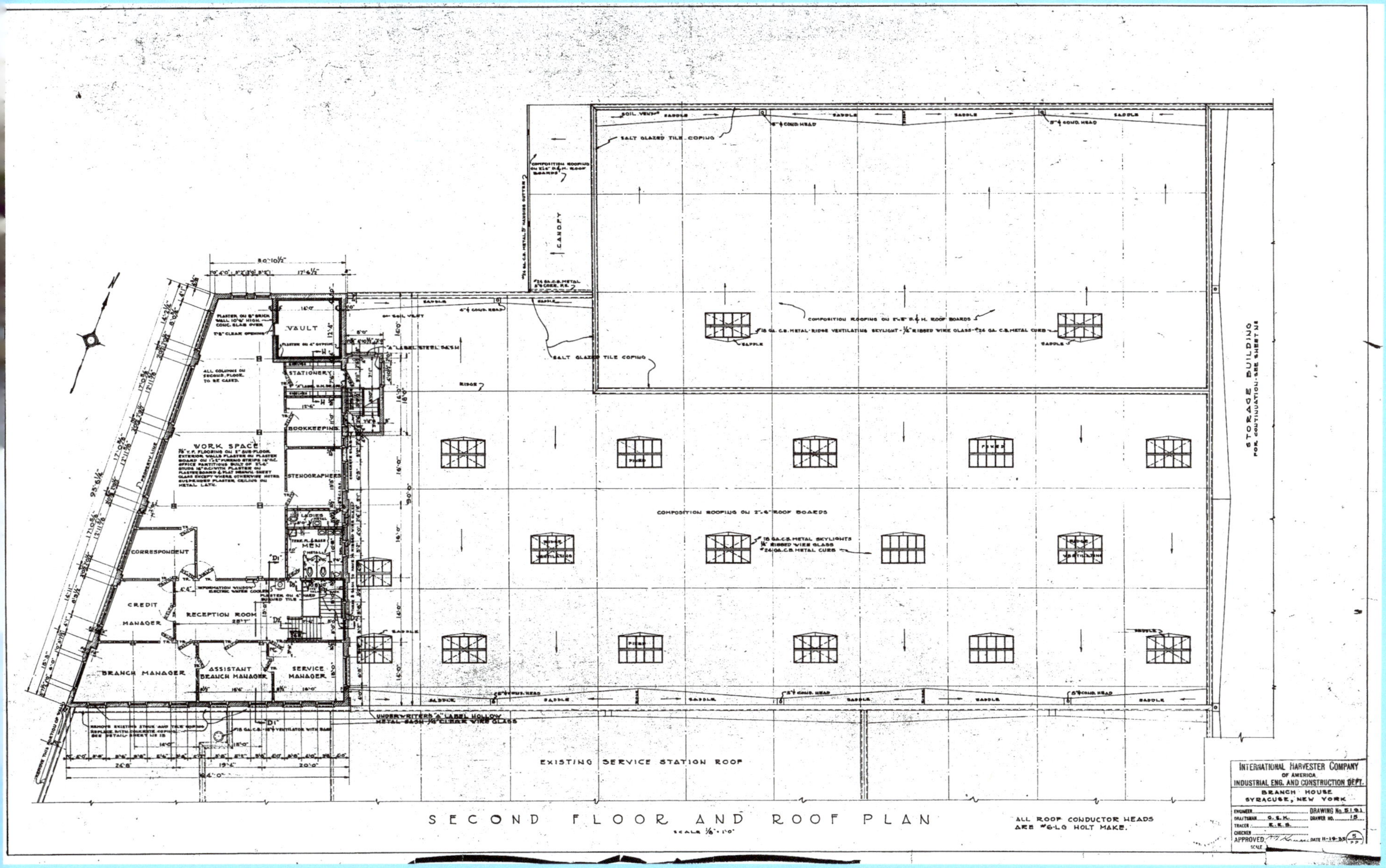

Second floor and roof plan of Service station and storage area.

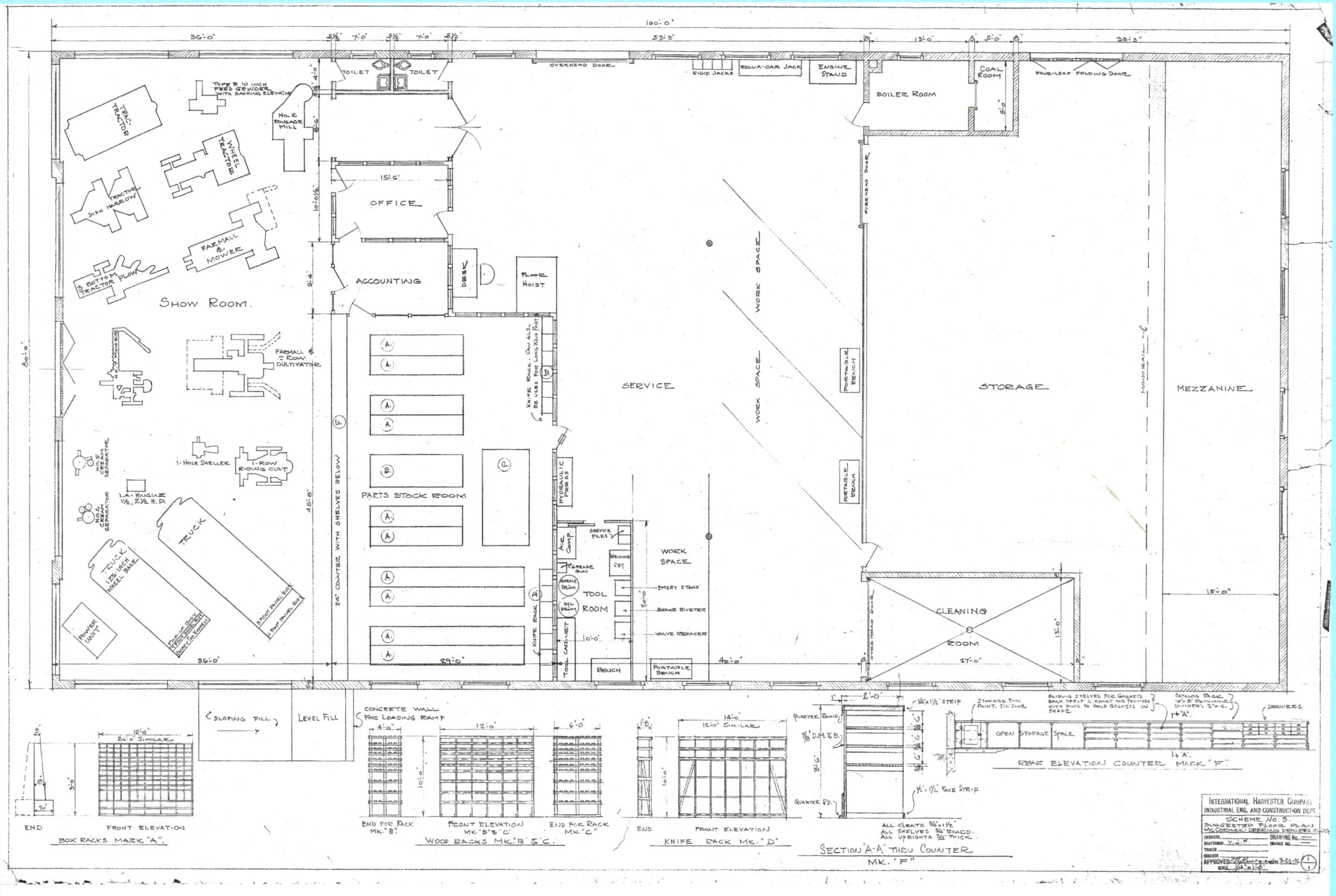

1923 layout of typical showroom display, for universal look throughout all dealerships, precursor to prototype building styles of 1945.

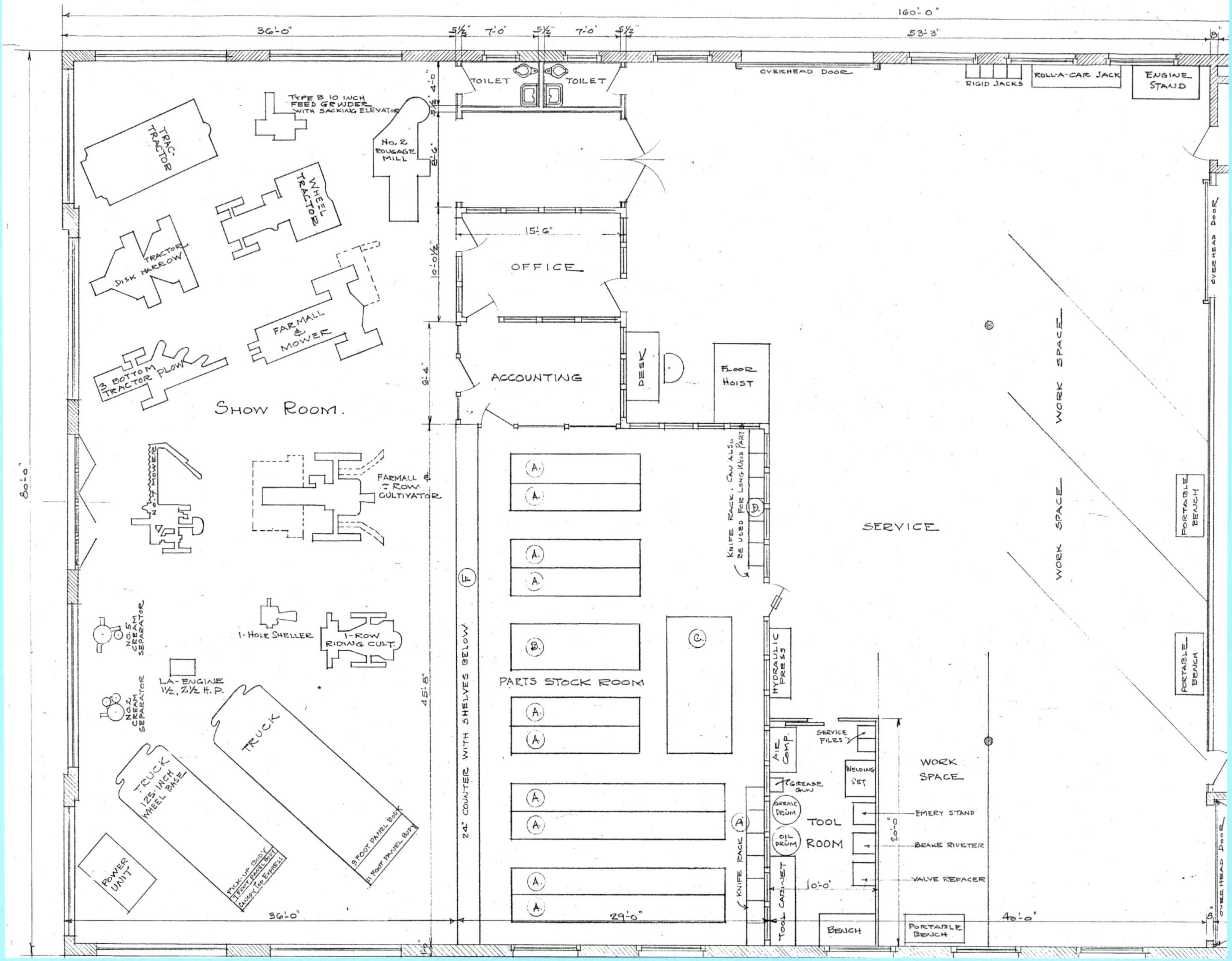

160'-0"
36'-0"
5½"
7'-0"
5½"
7'-0"
5½"
53'-3"
TOILET
TOILET
OVERHEAD DOOR
RIGID JACKS
ROLL-A-CAR JACK
ENGINE STAND
TYPE B 10 INCH FEED GRINDER WITH SACKING ELEVATOR
TRAC-TRACTOR
No. 2 ROUGAGE MILL
WHEEL TRACTOR
TRACTOR DISK HARROW
15'-6"
OFFICE
FARMALL & MOWER
3 BOTTOM TRACTOR PLOW
ACCOUNTING
DESK
FLOOR HOIST
SHOW ROOM.
MOWER
FARMALL & 2 ROW CULTIVATOR
KNIFE RACK, CAN ALSO BE USED FOR LONG WOOD PARTS
SERVICE
WORK SPACE
WORK SPACE
OVERHEAD DOOR
PORTABLE BENCH
PORTABLE BENCH
80'-0"
No. 5 CREAM SEPARATOR
1-HOLE SHELLER
1-ROW RIDING CULT.
LA-ENGINE 1½, 2½ H.P.
No. 2 CREAM SEPARATOR
24" COUNTER WITH SHELVES BELOW
45'-8"
PARTS STOCK ROOM
HYDRAULIC PRESS
TRUCK
TRUCK 125 INCH WHEEL BASE
AIR COMP.
SERVICE FILES
WELDING SET
WORK SPACE
GREASE GUN
GREASE DRUM
OIL DRUM
TOOL ROOM
EMERY STAND
BRAKE RIVETER
VALVE REFACER
KNIFE RACK
TOOL CABINET
10'-0"
20'-0"
POWER UNIT
PICK-UP BODY
9 FOOT PANEL BODY
11 FOOT PANEL BODY
36'-0"
29'-0"
BENCH
PORTABLE BENCH
40'-0"
OVERHEAD DOOR

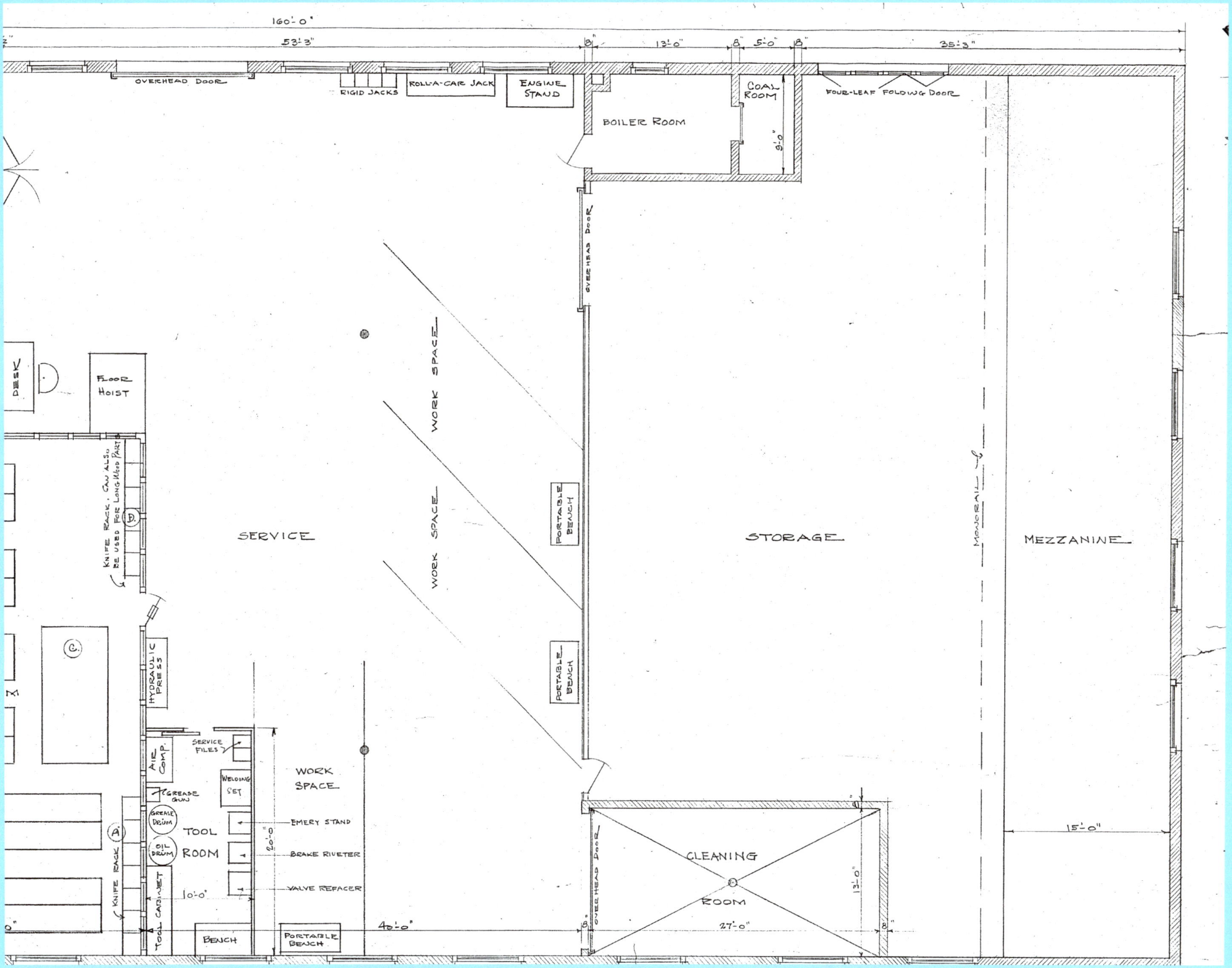

160'-0"
53'-3"
8"
13'-0"
8"
5'-0"
8"
35'-3"
OVERHEAD DOOR
RIGID JACKS
ROLL-A-CAR JACK
ENGINE STAND
BOILER ROOM
COAL ROOM
9'-0"
FOUR-LEAF FOLDING DOOR
OVER HEAD DOOR
DESK
FLOOR HOIST
WORK SPACE
WORK SPACE
KNIFE RACK, CAN ALSO BE USED FOR LONG WOOD PARTS
D.
SERVICE
PORTABLE BENCH
STORAGE
MONORAIL ℄
MEZZANINE
C.
HYDRAULIC PRESS
PORTABLE BENCH
AIR COMP.
SERVICE FILES
WORK SPACE
WELDING SET
GREASE GUN
GREASE DRUM
TOOL ROOM
OIL DRUM
A
KNIFE RACK
EMERY STAND
BRAKE RIVETER
VALVE REFACER
20'-0"
TOOL CABINET
10'-0"
BENCH
PORTABLE BENCH
40'-0"
OVER HEAD DOOR
CLEANING ROOM
13'-0"
27'-0"
8"
15'-0"

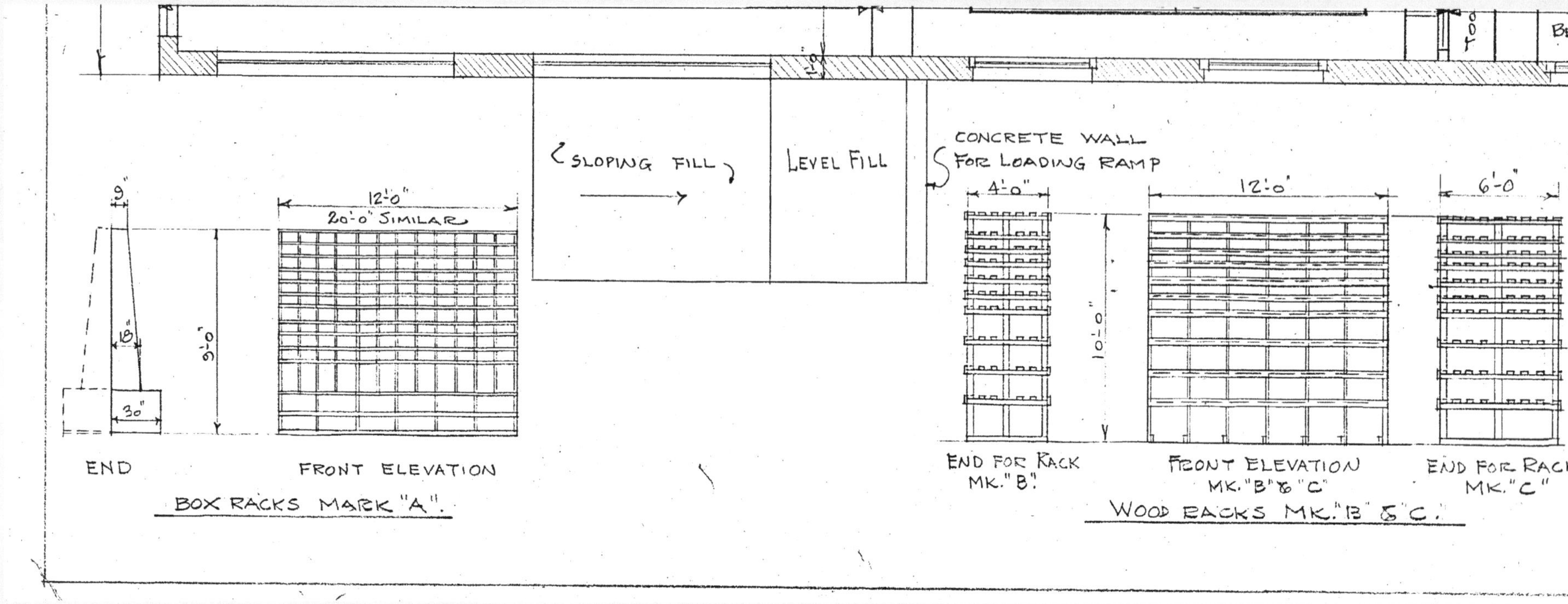

Through sections and elevations of parts bin, counter and racks with dimensions.

Prior page showing layout of floor plan for set-up and efficient working area to service both customers and repair shop.

These drawings vary slightly for the different square footage size of dealerships, with the basic layout staying the same.

New York plans from the Tomac Collection

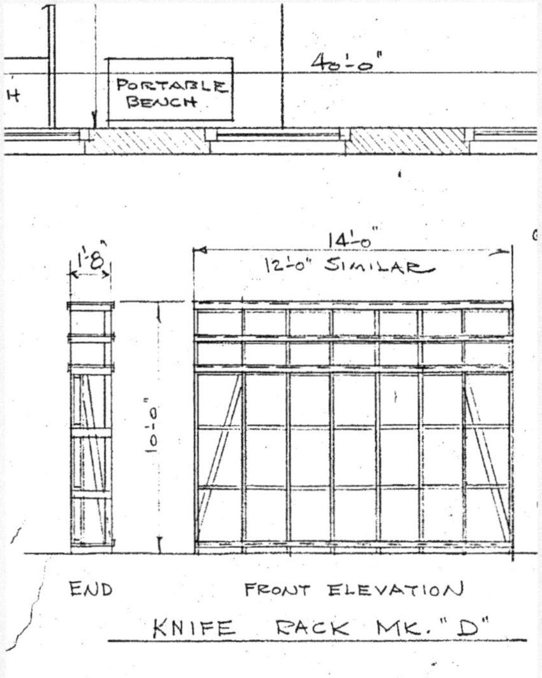
PORTABLE BENCH
40'-0"
14'-0"
12'-0" SIMILAR
1'-8"
10'-0"
END
FRONT ELEVATION
KNIFE RACK MK. "D"

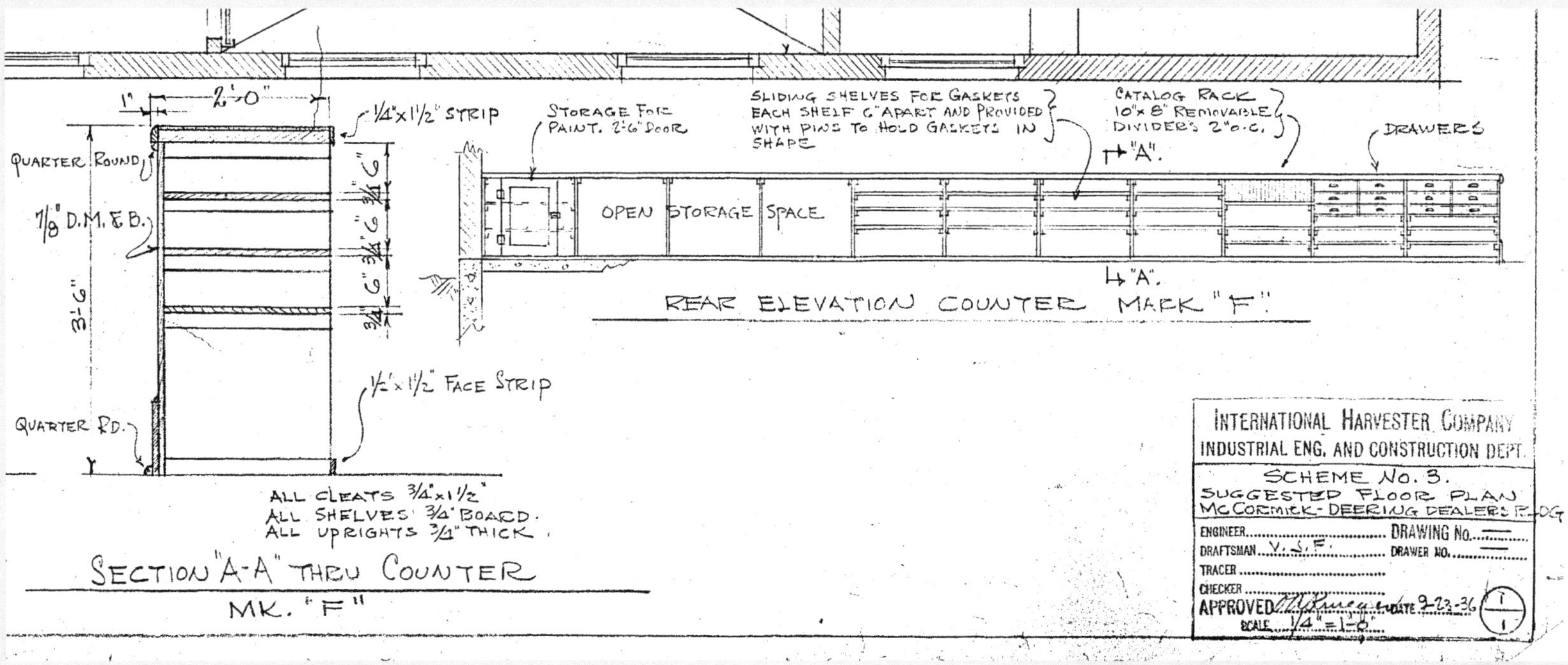
1"
2'-0"
1/4"x1 1/2" STRIP
QUARTER ROUND
7/8" D.M.&B.
3'-6"
QUARTER RD.
1/2"x1 1/2" FACE STRIP
ALL CLEATS 3/4"x1 1/2"
ALL SHELVES 3/4" BOARD.
ALL UPRIGHTS 3/4" THICK.
SECTION "A-A" THRU COUNTER
MK. "F"
STORAGE FOR PAINT. 2'-6" DOOR
OPEN STORAGE SPACE
SLIDING SHELVES FOR GASKETS EACH SHELF 6" APART AND PROVIDED WITH PINS TO HOLD GASKETS IN SHAPE
CATALOG RACK 10"x 8" REMOVABLE DIVIDERS 2" o.c.
DRAWERS
"A"
REAR ELEVATION COUNTER MARK "F"
INTERNATIONAL HARVESTER COMPANY
INDUSTRIAL ENG. AND CONSTRUCTION DEPT.
SCHEME No. 3.
SUGGESTED FLOOR PLAN
McCORMICK-DEERING DEALERS BLDG
ENGINEER
DRAWING No.
DRAFTSMAN V. J. F.
DRAWER NO.
TRACER
CHECKER
APPROVED
DATE 9-23-36
SCALE 1/4"=1'-0"

www.ingramcontent.com/pod-product-compliance
Lightning Source LLC
LaVergne TN
LVHW070819210926
848486LV00009B/370